Acta Neurochirurgica
Supplements

Research and Publishing in Neurosurgery

Edited by
Y. Kanpolat

Acta Neurochirurgica
Supplement 83

SpringerWienNewYork

Prof. Dr. Y. Kanpolat
Ankara, Turkey

Printing was supported by Turkish Academy of Science

TÜBA

Typesetting: Asco Typesetters, Hong Kong

Printed on acid-free and chlorine-free bleached paper
SPIN: 10875627

With 20 Figures

CIP-data applied for

ISSN 0065-1419
ISBN 978-3-211-83821-1 Springer-Verlag Wien New York

Preface

"Research and Publication" are phrases familiar to all scientists. Unfortunately, many scientists struggle with them on a trial-and-error basis, and there are no structured education programs providing information on standard methodologies. This is why the European Association of Neurosurgical Societies (EANS) Research Committee has developed a course on research and publication methodologies for residents in neurosurgery who have not yet completed training.

It is hoped that the Supplement will serve as an essential handbook for young neurosurgeons and researchers which provides the basic tools to guide their research and publication work, presents time-saving advice, and results in the most beneficial contributions in experimental and clinical research in neurosurgery and neuroscience. Such a collection of articles on this subject by different authors has previously not been compiled for publication; however, the material was presented as a training course under the auspices of the Research Committee of EANS, held in Antalya, Turkey, in 2001.

I would like to express my gratitude to the President of EANS, J. L. Antunes; and its Administrative Council; to H.-J. Reulen, Editor of Acta Neurochirurgica; the Turkish Academy of Science (TÜBA), which gave financial support for this publication; Springer Verlag, the publisher; and to the authors of this Supplement for their kind and generous support for completion of this project.

Y. Kanpolat

Contents

Listed in Current Contents

Acta Neurochir (2002) [Suppl] 83: 1–5

Significance of Research*

Y. Kanpolat

Department of Neurosurgery, Ankara University, School of Medicine, Ankara, Turkey

Summary

The development of the central nervous system (CNS) in humans is the most important factor differentiating the human being from other species. Intelligence is the most important result of evolution of the CNS in humans. However, human intelligence is not a static factor, having evolved during the historical progression of cultural factors and educational systems. Since the Middle Ages, universities have been the most dominant open-society institutions to regulate intellectual influences, share scientific knowledge and values, and promote research in natural sciences. Research is particularly important in contemporary neurosurgery and neurological sciences with regard to the development of new techniques and testing of the acquired knowledge of the CNS. Research activities in neuroscience and neurosurgery have especially involved anatomical and functional aspects. However, advances in molecular biology and genetics have been remarkably effective in research activities in neurosciences. Such progress will help us to understand nature, environment and humans. Of course, research is now a domain of method, organisation and financial gain. The dedication, heroism and creativity which used to be the driving forces behind scientific research are now diminishing in value and are a source of controversy.

Keywords: Evolution; science; research; creativity.

Research and publication are concepts we frequently talk about and use as the most reliable yardsticks in evaluating our scientific performance. However, I believe we must recognize that many of us have not acquired sound basics for implementing these concepts. In institutionalised research bodies, no doubt the basic methodologies for research and publication are well-known and properly taught. However, as one can see in the daily lives of our colleagues, there are those who do not exhibit a full understanding of these notions. Moreover, there is no established standard that governs research training in our day-to-day practice. We all agree that there is a necessity for research as part of the neurosurgical education. However, the best way to incorporate research into the neurosurgical training is still not clear. Should it be conducted prior to the neurosurgical education, simultaneously or immediately after? On the other hand, there are a limited number of publications or programs focussing on methodology of research and publication, which in itself is an important parameter of our training and practice [12].

The European Association of Neurosurgical Societies, "EANS", has successfully conducted various courses for *residents* in neurosurgery as joint training programs implemented by its Training Committee. The first course was effectively organized in Brussels from September 15 to 22, 1974, at the faculty of the Free University of Brussels. Professor Cohadon organised the first advanced seminar in Cortona, Italy, May 2–4, 1985, on "Traumatic Brain Oedema" [4]. In continuation of this effort, we decided to establish a training course on research and publication methodologies to aid orientation of our trainees in this direction. I am sure we, as trainers, have as much to learn from this as you, the trainees. If this series of courses achieves its goals, we hope to make it a part of the regular EANS training programs. We now publish the manuscripts of our first year's course, encouraged by the favourable responses received from the participants following the event.

The first course is devoted to the significance of research. In this context, I would like to start with a short review of our distant past.

The universe is estimated to be 12 to 14 billion years old. It is generally accepted that our planet earth separated from the sun – an insignificant star in a much

* This paper is dedicated to the memory of Late *Prof. Fausto Ianotti M.D.*, member of the EANS Research Committee who was a distinguished scientist, a person of excellent human values and a dear friend.

larger universe – about 4 to 6 billion years ago as a ball of fire in space [3, 13]. The sun is one of the major factors that support our existence but is also one of our greatest adversaries. Its ultraviolet radiation basically made emergence of life impossible [3]. However, filtering brought about by water vapour scattered into the earth's early atmosphere through volcanic eruptions somehow turned this harmful radiation into a life-supporting medium. On our world, the crust of which took hundreds of thousands of years to cool, life somehow started in its most primitive form, in a way no researcher has yet been able to affirm [3].

The simplest and the most meaningful experiment in this respect was conducted by Stanley Miller, a Chicago chemistry student, in 1953. He made an aqueous mixture of all the elements assumed to be present in the earth's crust and atmosphere at the initial stage of life's emergence, such as methane, ammonia and others. He put it in a reactor vessel and applied electrical sparks to simulate the lightning that presumably started the life-forming reaction on the primitive earth. At the end of 24 hours, three basic amino acids were found within the vessel: glycine, alanine and aspartate [3]. It is assumed that these proteins were the initiators of life within earth's otherwise lifeless nature. And we can assume that the first living cell, which began its adventure within the earth's muddy, salty waters billions of years ago, was formed as a result of chemical reactions between purins, sugars and porphyrins [3, 13].

Some of these cells developed roots in the muddy depths of lakes and creeks to become the predecessors of plants. Others, as a result of being tossed about by currents and due to an increasing need to move, developed arachnoid projections to become the predecessors of water crustaceans [13].

In time, life proliferated in the oceans, the number of species increased and plants formerly rooted to the sea and lake bottoms moved onto dry land. In the meantime, some fishlike species developed lungs and learned to breathe in order to live also on land. These were called amphibians due to their ability to live both on land and in water. In the early stages of life on land, there were oversized plants and giant reptiles. Most of the reptiles had a greater need for wings than for legs [13].

In time, the giant reptiles disappeared and were replaced by another animal species, later named mammals, that grew offspring inside their bodies, instead of laying eggs, and fed them with the secretions of their mammary glands. These mammals were our early predecessors. During their evolution, a species of mammals known as primates grew more proficient in finding food by using their well-developed upper limbs. A special kind of primate further developed in time so that it could better survive in nature. Its eyes shifted forward on the face, providing three-dimensional vision, and developments in bone structure and the nervous system allowed the primate to maintain an erect posture. Most distinctly, the primate evolved to acquire the advanced nervous system function that we now call *intelligence*. The most significant factor that differentiates humans from other living organisms is its ability to produce knowledge [3, 13].

The greatest concern of our early predecessors was finding food and shelter, tasks which occupied nearly all of their time. Many primitive humans settled in the deltas, where major rivers reached the sea, as these areas were more conducive to survival. There, rivers laid fertile lands eroded from other areas, and it was easier to cultivate plants for food. That left our early predecessors with time to lift their heads and observe each other and their environment. This phenomenon which helped to facilitate life on earth, is the second important milestone in the existence of humanity, in the context of research and exploration [13].

Some of these people were sensitive enough to observe the events taking place around them, view changes and sense by intuition their meanings and consequences. When they were able to derive the same results from certain events under the same conditions, they presented those to their contemporaries as basic facts about life and nature. Those were our first scientist ancestors.

There were others who were sensitive to their environment in a different way. Those people reflected their observations to works of art such as music, painting and sculptures. They were endowed with the special gift of being receptive to the vibrations of nature that others did not feel. Those were our artists. For example, Sophocles, around the 1st Century B.C., uncovered the psychological complex that governs the relations between mother and son as reflected in his play "King Oedipus". Two thousand years later the scientific basis of this psychological syndrome was presented to the world by Sigmund Freud. Science and art have since then continued to complement each other, like two sides of a leaf, as to how natural phenomena are perceived and evaluated.

There were also those who preferred to explain the existence of the human being within nature by supernatural allusions. For them, such allusions were incontestable facts and could not be changed. Those were our scholastically-oriented ancestors. They looked unfavourably upon those who contested their basic truths and wanted to question, review and change them. This conflict of ideas and ideals has not changed much throughout history.

Our predecessors could transfer their knowledge to the coming generations only through verbal communication. With the invention of writing, the accumulation and generation of knowledge increased at an extraordinary rate. Those who were in power and conscious of the importance of recorded knowledge acted as protectors of libraries that were the storehouses of knowledge, and science developed more rapidly in congregations and states where libraries existed. Eventually, those who wanted to acquire knowledge, gathered around those who had knowledge. What was important was the acquisition and sharing of knowledge. Hence, the word "*Universitas*" describes the union of people who possess knowledge and those who wish to learn. This is the basic idea behind present day universities [6, 13].

Curiosity and the urge for pioneering have been the basic motivation behind the realisation of inventions and discoveries, together with the desire to acquire wealth. For example, gold miners have been important leaders in enhancing discoveries in the area of chemistry. A thirst for power led to the development of more lethal weapons to provide military supremacy. Another motivation, which no doubt drives also those present here, is the desire to help humanity, a desire which has been effective in the development of the medical profession.

A marked acceleration in the emergence of knowledge, and of scientific and artistic expression was most apparent during the Renaissance. This was triggered by three developments. The invention of the printing press caused knowledge to be disseminated to masses. The discovery of the compass allowed for the navigation of large oceans and a widening of the horizons of mankind through the discovery of new lands. The invention of gunpowder showed the supremacy of technology over manpower in military campaigns [13].

In the same period, the first universities developed in the backyards of religious institutions. Foremost among those were the universities of Bologna and Paris. The University of Bologna, even though it was established in 1088, was independent of the church and the state [1, 6]. Its curricula were founded on freedom of thought, and its power came from its ability to employ freedom of thought. In the same era, a group of idealists emerged who were devoted to the idea of a unified Europe, joined by tolerance and co-operation between people and with Latin as the common language [1, 13]. We call these people the "Humanists," and Erasmus was one of the most important representatives of this ideal.

The University, in its early years, was an institution of professional education and training. It developed and proliferated across continental Europe around the common language of Latin. Indeed, the foundation of today's European Union was laid then. As we all know, today's EU is founded on the concepts of internal independence, willing cooperation, goodwill, and tolerance between the member nations and their institutions [1, 5].

In the 18th century, the concept of conducting research in universities was launched by Baron von Humboldt, a German idealist. This idea was most widely accepted by American and European universities, and they were first to excel as research institutions [5].

Ever since, new concepts and products emerging as a result of scientific research in the universities have penetrated industries, initiating the modern concept of industry-university cooperation [5]. This eventually resulted in a closer collaboration leading to the establishment over university campuses of industrial technoparks. That opened the way to the more intimate kind of university-industry-community relationship, contributing to the welfare of nations. As an example, one can show the great innovation and research financing potential that exists in the U.S. California is now one of the wealthiest states in the U.S., and this region receives an *extraordinary* share of wealth as a result of the high-tech products produced there. The roots of this high-tech sector are firmly embedded in the soil of such prominent universities as Stanford, Caltech, Berkeley, and the University of Southern California [5].

Microsoft, one of the world famous U.S. high-tech companies, has attained today a market value of roughly $369 billion. The company, founded in 1975, employs a total workforce of around 40,000. On the other hand, 10 companies, which are marketing and producing high-technology, account for half of the 20 most important companies in the world, considering their market value [9]. This illustrates a trend in the modern world that innovating intellectual power –

more than labour, natural resources and capital – is the key to success.

However, it is also extremely important to achieve the proper balance between the main functions of the university as an institution of education and research versus an institution that augments the innovative capacity of the industries. I am sure the proper instrument to achieve this balance in favour of protecting the basic functions of university is the scientific intellectual reflex that exists inherently in the concept of "university".

Getting back to the practice of neurosurgery, Hoff likens the practice to a three-legged stool consisting of patient care, education and research [6]. Speaking of research, as our main concern here, one can see that the basic research models in neurosurgery have developed in three fundamental stages.

Initially, research is inclined to explain the morphology or anatomy of the nervous system. Important contributions in this area of painters and sculptors such as Leonardo da Vinci, Michelangelo and others have in time been replaced by the practice of the anatomists, especially Andrea Vesalius, who performed in spectacular rituals [2, 7, 8]. Modern practice in anatomy, with the aid of the surgical microscope and computers, has been greatly instrumental in developing our surgical skills and orientation.

Function-based research, although initiated at the time of Galen, has been substantially developed by the efforts of Descartes, Willis, Galvani, Sherrington, Claude Bernard and Pavlov. With the emergence of methods of electrophysiology, our function-based understanding of the nervous system has increased exponentially [2, 8].

Molecular-based research is the product of a more recent past. Modern trends have enabled us to examine cellular function in view of chemical, physical, genetic and immunological approaches. Recent developments in molecular biology have opened the way to the development of a new scientific understanding in neurosurgery, which we now call "molecular pathology". Molecular pathology enables us to trace the molecular origins of neurological disorders and provides us with the means of dealing with these disorders using molecular scale techniques, both at the operative and clinical stages as well as at the preventive stage. Further research with the use of molecular biology techniques will also provide thorough understanding of normal and abnormal properties of cells, tissues, organs and systems at molecular disease, which will contribute to the progress in biomedical science [11]. It is reasonable to think that molecular neurosurgery will very soon become an established practice in our profession. Obviously, in this climate of busy research activity at all levels, our understanding as well as realising the extent of our ignorance will increase.

Modern technology in step with the trend of development in neurosurgery has led to a constant bombardment of information and data concerning new products and techniques, forcing us into a life style that is greatly governed by high-tech or, more appropriately, by company-tech. It is up to true researchers to decide how well this high-tech approach will answer the needs of the society, the human being and the medical practitioner.

Nowadays, this data overload emerges as the fourth leg of the neurosurgical stool. Dizzying developments in information technology, as well as rapid changes in technological products, have resulted in an information explosion. Expensive tools produced by the industries today have no doubt improved our surgical techniques and results. However these high-tech products are prohibitively expensive and are conducive to a way of life that is predominantly controlled by business. Our age is the age of information technology but it is also the age of companies. With our growing knowledge, our problems will also increase.

The solution to our problems will be found, again, only by the use of human intellect. The human intellect, through its ability to merge knowledge, philosophy and art, carries the inherent capability to provide a paradise on earth for all humanity. On the other hand, to be overwhelmed by the possibilities offered by modern science and technology and indulge in its exploitation in disregard of ethics and the normal functioning of nature, can easily lead scientific achievement to a position that may be extremely detrimental to humanity.

One cannot be a scientist only by storing data, attaining data and using the existing accumulation of data, it involves innovation and intellectual progress. Being a researcher and a scientist is not just a profession. As the famous Turkish mathematician Cahit Arf once said, "*Science is a way of life*". It is the ability to sense the realities around us that are hidden to others, by instincts. As Isaac Newton once said: "*I know not what I may appear to the world, but to myself I appear to have been only like a boy playing on the sea-shore, and diverting myself in now and then finding a smoother or a prettier shell than ordinary, while the great ocean of truth lay all undiscovered before me" [10]*. In reality, being a researcher is inherent in the ability of keeping alive in ourselves that curious boy asking questions by the ocean side.

Today's mass media and even the system of education force us to standardise this curious boy, diminish social differences between our people and make a prototype of local cultures. I. I. Rabi, the Nobel Physics Laureate in 1991, attributes his abilities to his mother's very special approach to him when he was a grade school student. When he returned from school every day, his mother would ask not "Did you learn something today?" but *"Did you ask a good question today?"* It is this approach and intuition that allow one to penetrate what is apparent and see the truth behind it [10]. This approach and intuitive thinking, when supported by a certain methodology, accelerate success in research. It is no doubt important to produce individual researchers. But what is more important is to be able to make research a system and even more a tradition, that lives on within the scientific community.

The point is that we must sustain our earnestness and a child's curiosity when involved in science and research. I am convinced that, under the assault of the abundance of information and data, the ability to preserve our curiosity and our humanity are of foremost importance when approaching scientific research.

Acknowledgments

I would like to extend my thanks to my friend *Ender Arkun* for his contributions to the text, my friend *Murat Tuzcu M.D.* for his review and comments and to Mrs. Corine Kora for editing this article.

References

1. Brihaye J (1989) The European Spirit and the European Association of Neurological Societies. Acta Neurochir (Wien) 99: 1–1
2. Clarke E, O'Malley CD (1996) The Human brain and spinal cord. Norman Publishing, San Francisco
3. Ditfurth HV (1974) Im Anfang war der Wasserstoff. Hoffman und Campe Verlag, pp 4–42
4. Guidetti B (1987) European courses of neurosurgery and other training and postgraduate education activities of the European association of neurosurgical societies. Acta Neurochir (Wien) 85: 115–128
5. Guruz K (2001) Dunyada ve Turkiyede Yuksek Ogretım.- OSYM Yayinlari Ankara, pp 1–182
6. Hoff JT (1994) Toward better balance. The 1994 Presidential Address J Neurosurg 81: 651–655
7. Keele DK, Roberts J (1978) Leonardo da Vinci. Anatomical Drawing from Royal Library, Windsor Castle. Griffin Press Ltd, Netley
8. Lows ER, Udvarhelyi GB (1998) The genesis of neuroscience by A Earl Walker. AANS Publications
9. Matlak C (2001) The global 1000. Business Week, July 2001, pp 46–81
10. Meyer MA (1995) Science creativity and serendıpity. AJR 165: 755–776
11. Santis G, Evans GT (1999) Molecular biology for critical care physician, part 1. Terminology and technology. Care Med 27(4): 825–831
12. Reulen HJ (1997) When should residents be exposed to research? Acta Neurochir (Wien) [Suppl] 69: 106–110
13. Van Loon HW (1957) The Story of Mankind. The pocket library, New York, pp 3–25

Correspondence: Prof. Dr. Yucel Kanpolat, Inkilap Sokak 24/2, Kizilay, 06640 Ankara, Turkey.

Acta Neurochir (2002) [Suppl] 83: 7–9

How to Construct a Research Project

T. Dalkara

Department of Neurology, Faculty of Medicine and, Institute of Neurological Sciences and Psychiatry, Hacettepe University, Ankara, Turkey

Summary

A research proposal should start with defining its aims. A synopsis of the literature and observations that the hypothesis is grounded should be given together with relevant references. The significance of the project, expected contributions to the field should also be indicated. If available, preliminary data enhance the impact of the proposal. A detailed description of the methods and subjects to be used is an important part of the proposal. What data are to be collected, the method of collecting data, and selection criteria for subjects should be indicated. Variables and how the measurements will be taken, have to be defined precisely. Data processing and analysis tools should be described. Additionally, the expected time table, project cost, the ethical and legal issues should be included in the proposal. It should be kept in mind that a scientist's primary responsibility is to create conditions such that the hypothesis can be tested objectively rather than proving his hypothesis.

Keywords: Research project; hypothesis; experimental design; data collection.

We can simply classify research projects as basic, applied or developmental. In basic research, the aim is to further the existing body of knowledge in any field. The applied research (eg. clinical research) also aims at contributing to the theory but it also has a practical goal: the consequences of the new information obtained must be applicable to solving an existing problem. Developmental projects do not aim to contribute to the theory but have a purely practical goal such as devising a better surgical approach or instrument. Some studies are descriptive; they collect data describing the subjects under study (eg. population demographics). Other studies use analytical tools to understand the relationships between several variables (eg. diet and blood pressure) and elucidate mechanisms of some processes (eg. pathophysiology of diseases).

An Investigator's Primary Responsibility Is to Create Conditions Such That the Hypothesis Can Objectively Be Tested

Scientific progress has a spiral-like course: Every research project begins with an observation and sets out to answer a question or to solve a problem, however, may lead to new observations that prompt new questions while providing answers to the original question (Fig. 1). Not infrequently, the answers found come about as a total surprise. An orthodox belief in the original hypothesis and current theories may blind researchers and cause them to miss these unexpected answers, which are an exciting part of the scientific endeavour. An investigator specifies the conditions in a way that his hypothesis can be tested, performs the experiments and collects the data in accordance with the established protocol. The results may support the hypothesis or may require revision of the original idea or disprove it. As long as the research project is prop-

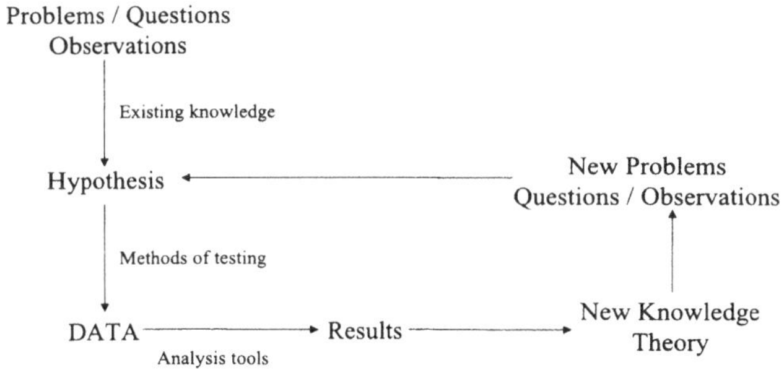

Fig. 1

erly designed, the conclusions drawn will contribute to the existing knowledge whether or not they support the original hypothesis. Therefore, a scientist's primary responsibility is to create conditions such that the hypothesis can be tested objectively rather than proving his hypothesis [2].

To achieve objectivity is one of the most difficult goals of scientific research. The transformation of human thinking from subjectivity to objectivity was one of the major leaps in human history. Elimination of subjective tendencies in a scientist's attitude requires training and experience. However, human beings are inherently vulnerable to subjectivity, and the desired level of objectivity can only be obtained collectively by publishing our results. Even in the publication medium, an ideal level of objectivity can sometimes be reached after long debates. Ideas and hypothesis are easy to propose. They usually sound great to us. However, a hypothesis can be turned into a piece of scientific information only after passing through the painstaking process of objective evaluation. Therefore, the hypothesis proposed must be testable [2, 4]. No matter how bright it sounds, if we do not have the means to test it, a hypothesis remains a speculation until the means for testing become available.

Preparing a Research Protocol

A research proposal should start by clearly stating its aims (Table 1) [2]. A synopsis of the literature and observations that the hypothesis is grounded should be given together with relevant references. Review of the literature is expected to be limited to the specified aims only. The significance of the project, expected contributions to the field should be indicated. If available, preliminary data enhances the impact of the proposal, therefore should be described after the background and significance of the project have been introduced. Next, we should indicate how the hypothesis can be tested. A detailed description of the methods and subjects to be used is an important part of the proposal. Experimental conditions should be completely outlined, strengths and possible limitations of methods should be discussed. Researchers should make it clear that they have the required expertise in the techniques to be used and are familiar with limitations and pitfalls. Improper or erroneous use of methods can lead to misleading conclusions. Therefore, execution of test methods represent the most critical step of research projects. Junior investigators may think that established methods are flawless and hence disregard the problems inherent in methodology. Clearly, every investigator is expected to be critical of the quality and reproducibility of results.

Table 1. *Preparing a Research Protocol*

1) What are the aims of the study?
2) What is already known about the problem?
3) What are the preliminary data?
4) What design will be used in collecting data?
5) How are the subjects of the study to be chosen?
6) What data are to be collected, and why?
7) How are the variables to be defined and measured?
8) How are the data to be collected and the measurements to be made?
9) How will the data be processed and analysed?
10) What is the expected time table for the study?
11) What will the project cost?
12) What are the ethical and legal issues?

Depending on the way of data collection, research projects can be designed as prospective or retrospective studies [3]. In prospective studies, data are collected after the research project has started, whereas a retrospective study uses data already collected such as data from patient files. Although retrospective studies cost less effort and time, prospective studies are to be preferred because high standards for selection of subjects, methods and means of data collection cannot be assured for retrospectively collected data. It should be kept in mind that statistical tests assume that the data generated come from a group of subjects carefully selected to represent the population under study. They also assume that all other variables except those under study were similarly controlled for each subject. Therefore, a detailed description of the subjects, sampling techniques and of experimental groups and conditions based on which the data will be collected is essential for a reliable data analysis. An experiment or a clinical trial may generate a wide range of data. Therefore, what data will be collected, why and how these data will be collected should also be addressed [2, 4]. A clear definition of each variable in clinical studies, well described inclusion and exclusion criteria and randomization method for patient selection are required [3]. Non-randomised subject selection can introduce bias or unintended variables to the study.

The number of subjects to be studied should be large enough to give reliable results [1, 3, 4]. If the subjects act as their own control, the significant differences can be detected by a small number of subjects. Using an independent control group increases the number of subjects required, which is inversely proportional to the magnitude of the expected difference between groups. Sample size calculation techniques can help determine the required number of subjects in each group for a predetermined level of statistical significance, if data variation and expected difference between groups could be estimated. Ideally, the control group is subject to the same conditions as the test group and receives a placebo or is sham operated. Placebo and sham operation can have significant effects on patients and experimental animals, hence, the importance of including them into the study should not be underestimated.

When possible, researchers should be blinded during data collection and processing [1]. Although open-label studies are less complicated to perform, blinded studies increase the reliability of the data, especially when objective standards of data collection, measurement and evaluation are difficult to achieve. In studies on human subjects, it is furthermore desirable that the subject is also blinded (double-blind study). Crossing over treatments the groups are receiving after a certain trial period (e.g. group 1 switches from A to treatment B whereas group 2 from B to A) increases the reliability of the data, particularly of those based on patient reports (double-blind crossover study) [1].

Quantification of the data also represents an important step. Every investigator must critically evaluate the methods of measurement to obtain correct readings at an acceptable level of precision [2, 4]. The measuring devices, how the measurements will be made, and the level of precision should be described in detail. After making sure that a reliable body of data will be produced, the next step is to choose the correct methods of data evaluation. The statistical tests to be used should be clearly specified in the proposal. Comparisons of variables not defined in the proposal should be avoided although they may be showing statistically significant results. Statistics should be corrected for multiple comparisons because they may yield significant results by mere chance simply due to an increased number of comparisons. There are striking examples in the medical literature of some irreproducible results because those studies had not been designed to test the reported findings but comparisons between several variables in the collected data showed statistically significant results. Therefore, positive results obtained from unintended measurements should encourage new research projects to specifically test these observations in future studies rather than lead to premature conclusions.

An estimate of the duration and cost of the research projects is also required and give hints to the reviewer about the investigator's research experience [2]. A realistic time table should be added. The time table should specify in detail the distribution of both the work load and expenses over the entire period of the project. How the results of the project will be reported or disseminated and expected practical applications should be indicated. Finally, research projects should observe the ethical regulations with regard to use of human subjects or experimental animals and should include all the necessary documents. Similarly, legal documents regarding conflict of interest, patent issues, patient's rights etc should be provided.

Pilot Studies

A modification in either methods, subjects, data quantification or statistical tests is not desirable during or at the end of a project [2]. Such changes, although sometimes inevitable, decrease the reliability of results. To avoid this, one may start with pilot experiments performed with a smaller number of subjects. Such experiments may decrease the need for future revisions and avoid possible regrets in the end. Not infrequently, researchers wish to have the experiments, subject selection criteria, analysis tools, etc designed in a somewhat different manner than the original ones. Pilot experiments may decrease this risk.

References

1. Jones B, Kenward MG (1989) Design and analysis of crossed over trials. Chapman and Hall, London
2. Lock S (1985) How to do it, 2nd edn. Brit Med Ass
3. Matthews JNS (2000) An introduction to randomised controlled clinical trials. Arnold, London
4. Polgar S, Thomas SA (2001) Introduction to research in health sciences. Churchill Livingstone, London

Correspondence: Turgay Dalkara M.D., Ph.D., Department of Neurology, Hacettepe University Hospitals, Ankara, 06100, Turkey.

Acta Neurochir (2002) [Suppl] 83: 11–15

Research Ethics and Scientific Misconduct in Biomedical Research

E. Kansu and **S. Ruacan**

Hacettepe University, Institute of Oncology, Ankara, Turkey

Summary

Scientists have the responsibility of judging what is best for the patient and the optimal conditions for the conduct of the study. All physicians should ensure that research they participate in is ethically conducted. Every clinician should learn and receive training in the responsible conduct of research and publication, and each project must be reviewed by an institutional review committee.

Scientific misconduct is defined as any practice that deviates from those accepted by the scientific community and ultimately damages the integrity of the research process. "Sloppy Research" and "Scientific Fraud" include activities which can violate science, records and publication. Sloppy research is due to absence of appropriate training in research discipline and methodologies. In contrast, scientific fraud is defined as deliberate action during application, performance of research, and publication. It includes piracy, plagiarism and fraud.

Research institutions should adopt rules and regulations to respond to allegations, start investigational operations and perform appropriate sanctions.

Keywords: Research Ethics; Scientific Misconduct; Scientific Fraud.

Scientific research is based upon values such as integrity, honesty, trust as well as respect for academic, scientific and intellectual achievements. Integrity of the research study reflects the adherence of scientists to honest and reproducible methods in proposing, performing, evaluating and reporting research. Ethics are the branch of philosophy dealing with the concepts of honor, truthfulness, moral values, objectivity, honesty and integrity [2, 10, 11].

Scientists for centuries have been depending on each other and the rules of their community to protect the honest and objective measures during the research process. The concept of integrity and ethics is at the heart of research practice, clinical practice and publication issues that concern young investigators, chairmen, academic staff, and editors. Each individual's personal integrity in an academic institution as a researcher or a staff-member leads to the integrity and ethics of their institutions [2, 11, 15].

All the senior staff and lecturers have the responsibility for teaching ethical principles to new trainees in the institutions. There is an important need to teach junior scientists and students about research and publication ethics at a very early stage of their education [16, 20].

There are six essential components that have an effect on the results of biomedical research [15]:

a) Investigator: Person who is responsible for conducting the study and if successful research study leads to career advancement, promotion and possibly financial gains.
b) Employer/Institution: The place where investigator performs his/her research studies. The institution may gain credibility and profitability by the success of the researcher.
c) Sponsor of the research
d) Patient (in clinical research studies)
e) Scientific Community: Group which needs reliable scientific information
f) Public: Community who pays for biomedical research through taxes and donations.

Scientists who do clinical research involving their patients must have the responsibility of judging between what is best for the patient and what is optimal for the conduct of the study [15, 20]. In any case, physicians as researchers must consider their primary role as care-supplier "first" and as investigators "second". All physicians should ensure that research they participate in is ethically conducted [19, 20, 23].

Each investigator must be aware of the following issues before, during and after the research process [16, 20, 23]:

a. Validity

- A study is scientifically valid if it answers the questions that it asks.
- It should have a large enough number of subjects to provide statistically valid results.
- The techniques employed should be reliable, reproducible and sufficient to test the hypothesis.
- The study should not risk human subjects during the process.

b. Value

Valuable research has to designed to produce knowledge that ultimately proves to be important, reproducible, productive and contributory. The scientific community and the peers have to benefit from the results [11, 20].

c. Ethical Issues

Each research study must be reviewed by an interdisciplinary review committee or has to meet strictly outlined criteria for review [20].

Ethical committees should review several aspects of a proposed study including its risks, benefits, consent forms, the importance and impact of the new information to be gained and the confidentiality issues [19, 20, 23, 25].

Each human study dealing with patients and volunteer control subjects should be submitted to an ethical review committee. Investigators without access to an ethical committee may wish to contact to nearest academic medical center willing to review the protocol [16, 25].

d. Compensation

Payments should commensurate with the time and effort spent and the expenses incurred in recruitment [16].

e. Authorship

Authorized authorship requires involvement in developing a study's conception and design, analyzing, performing and interpreting results, drafting or revising the manuscript's intellectual content and approving the final text [3, 22, 24, 25].

Clinicians who are interested in contributing to research should spend some time learning about the responsible conduct of research and should receive training and advice from experienced senior investigators, because ethical conduct and scientific research demands careful consideration, planning and attention to detail [16, 20].

Scientific Misconduct

Scientific misconduct is defined by the U.S. Public Health Service as "any practices that seriously deviate from those that are commonly accepted within the scientific community for proposing, conducting or reporting research and ultimately damage the integrity of the research process" [20, 23]. However "questionable research practices or sloppy research" should include research practices or actions that violate the values of research process due to inadequate training and supervision [6, 20, 22, 24].

"Sloppy research" or questionable research practices include activities which can violate traditions of science, waste time and resources. These activities include failing to retain significant research data, keeping inadequate research records, utilizing inappropriate statistical methods or publishing preliminary research data without peer-review or validation. Most common cause of "Sloppy Research" is the absence of appropriate training in research and research methodologies [6, 20, 22].

Although some scientists have suggested that incidents of misconduct in research and publications are underreported, estimates given in U.S. goverment studies have been low. The Office of Scientific Integrity (OSI) in the Public Health Service in Washington, D.C. found evidence of misconduct in fewer than 20 cases from March 1989 through March 1991 [3, 20, 22]. Infrequent case disclosures of scientific misconduct still raise important concerns and questions among grant financers, scientists, institutions, public and media. Scientific misconduct has to be taken into account as a serious matter. The Royal College of Physicians classifies "scientific misconduct" as Piracy, Plagiarism and Fraud [3, 6, 22].

a. Piracy

Deliberate exploitation of ideas, work, text or other materials from another person (s) without acknowledgement [3, 20].

b. Plagiarism

Using or copying the ideas, data, text, words or illustrations of another person without giving appropriate credit (citation), without permission or acknowledgement, of the original author's consent [3, 20].

c. Fraud

- Fabrication: Deliberately making up data or results not obtained from the research, and report or publish these data.
- Falsification: Deliberately changing or modifying data or results, research materials, equipment, research records which leads to different results. These acts also include failure to perform research on-schedule, improper reporting of the status of subjects, selective reporting of primary data, using faulty statistical methodologies and inappropriate authorship practices.

Scientific misconduct and "fraud" includes mainly deliberate work or action during the application, performance of research studies, presentation and/or publication process [3, 6, 18, 20, 21, 22, 24].

The other typical examples of scientific misconduct in publications are as follows [3, 20, 22]:

Duplicate Publication

Republication of the same information, either as an entire paper or as information, of smaller dimensions than a complete paper. It also reflects repetitive publication representing objectively and more accurately the appearance of the same information two or more than two times. The definition covers the republication of an entire paper or a closely similar version representing the same data of the research.

Divided Publication (Least Publishable Unit, LPU)

The information from a single study is divided for publication into two or more papers. This practice has also been called "salami science" because it represents, in effect, the slicing up of a single study into several fractions.

Unauthorized Authorship Practices

Deletion or failure to acknowledge the names of those who significantly contributed to the research/article or inclusion of new authors for their minor contributions, without the written consent of all researchers and co-authors or changing the order of authors.

Reasons of Misconduct

The reasons for scientific misconduct or fraud [3, 4, 18, 20, 21] are as follows:

a. Inadequate or lack of research discipline, training and research ethics.
b. Pressure to publish by the department and/or institution
c. Personal ambition for rapid career advancement and academic promotion
d. The desire for acknowledgement
e. Financial gain
f. Psychiatric illness

In some cases the factors responsible for scientific fraud are still not fully understood [1, 4, 18].

Prevention of Scientific Fraud and Misconduct

There are mainly three categories of approaches for prevention of scientific misconduct [3, 8, 20]:

- Education, training and establishment of ethical standards
- Encouraging practices to reduce the pressures predisposing to misconduct
- Investigation of alleged misconduct

Education Training and Standards

All educators and mentors in scientific research and clinical practice have the responsibility for setting highest standards for "Good Research Management (GRM)". These standards represent an important means by which research institutions can show and support ethical research practices [5, 18, 20]. If there are certain standards and guidelines designed for ethical research conduct, the mentors will have a better ability to inform their trainees and members about what constitutes questionable practices or misconduct in science [5, 7, 17, 18].

There are certain principles which should be adopted by scientists in academic research environment [13, 17].

a. The numbers supervised by each senior investigator must be limited to ensure that each trainee receives adequate individual attention in regard to conduct of the project, data analysis and principles of publication.
b. Research data must be preserved in permanent "research data notebooks" all volumes should remain available and accessible for review, minimum for five years, preferably for ten years. All institutions have the responsibility of preserving research data securely for on-site visits and review processes.
c. In each study, expert statistical advice and consultation should be obtained, data should be fully reviewed prior to publication, relevant citations should be acknowledged and detailed experimental or clinical information should be given in the final manuscript [14].

In summary, research institutions should prepare guidelines to establish and maintain high standards for the conduct of research [9, 13]. Research conduct guidelines are generally aimed at promoting honest and objective conduct of research and reducing the incidence of misconduct in science.

Reducing the Pressures Predisposing to Scientific Misconduct

Reducing Unnecessary Pressure to Publish

The research project should be selected to meet the training and career needs of the young scientist so that results can be expected within a reasonable time. Young researchers should not generally be engaged on a project which is speculative. Researchers and junior staff should be trained that quality rather than quantity has a greater value in research and academic performance [1, 4, 12, 18].

Reducing Financial Pressures

Financial interests in biomedical research may include salary, consultation fee, licensing agreements, honoraria, research grants and financial support for meetings and travel. Research institutions may prohibit certain financial relationships and they should have more strict rules regarding the management of conflict. None of the sponsors can be assured that a grant or contract will yield discoveries or a specific "desired outcome" [13, 18, 21].

Investigating the Alleged Malpractice

All the research institutions and academic agencies should establish policies and regulatory guidelines to ensure a prompt reaction to allegations of misconduct in science [20].

Institutional process must include the following steps:

- Inquiry
- Investigation
- Reporting
- Implementation of Sanctions and Punishments

Research institutions should adopt policies which will ensure appropriate and prompt responses to allegations of misconduct in science. Both accused and accuser are entitled to anonymity during the early phases of misconduct-in-science investigations.

During the inquiry phase the institution must determine whether there is sufficient substance to the allegation to warrant an investigation. The whole process should proceed in a confidential manner and the alleged scientist should not be personally damaged. All inquiries should be conducted within the institution. The whole inquiry should be arranged as to ensure the confidentiality of process and investigations should be completed as rapidly as possible, preferably within 3 months.

Variety of sanctions are possible. In cases involving falsified or fabricated research, many journals have the policy that the relevant paper in its entirety will be retracted from the world literature. Staff who is the subject of proven scientific misconduct or fraud should resign from the research institution. All steps should be taken to ensure that the person will not transfer his/her research studies to another institution. Federal agencies in the United States and in Europe (NIH, NCI etc.) have occasionally demanded the return of governmental funds granted for research which had been misconducted.

Investigators who are convicted for research ethics violation should be notified in writing to the institutions where they are registered and to the journals concerned.

References

1. Altman L, Melcher L (1983) Fraud in science. Brit Med J 286: 2003–2006
2. American College of Physicians (1998) Ethics Manual 4th edn. Ann Intern Med 128: 576–594

3. Andersen D, Attrup L, Axelsen N, Riis P (1992) Scientific dishonesty and good clinical practice. Published by the Danish Medical Research Council
4. Angell M (1986) Publish or perish: a proposal. Ann Int Med 104: 261–262
5. Angell M, Relman AS (1988) Fraud in biomedical research: a time for congressional restraint. NEJM 318: 1462–1463
6. Culver CM, Clouser KD, Gert B, Brody H, Fletcher J, Jonsen A, Kopelman L, Lynn J, Siegler M, Wikler D (1985) Special report: basic curricular goals in medical ethics. NEJM 312: 253–256
7. Darsee Affair (1983) Retractions. New Engl J Med 308: 1400
8. Department of Health and Human Services (DHHS) (1991) First annual report: scientific misconduct investigations reviewed by office scientific integrity review, March 1989–Dec 1990. Office of Scientific Integrity Review, Washington DC, USA
9. Erkut S, Mokros JR (1984) Professors as "models" and "mentors" for college students. Am Educ Res J 21: 399–417
10. Ethical Issues in Biomedical Publication (2000) In: Jones AH, McLellan F (eds) The Johns Hopkins Univ Press
11. Freedman B (1987) Scientific value and validity as ethical requirements for research: a proposed explication. IRB 9: 7–10
12. Gunsalus CK (1999) The Baltimore case: a trial of politics, science and character. In: Kevles DJ (ed) (Book review). NEJM 340: 242
13. Harvard Medical School (1988) Guidelines for investigators in scientific research. Cambridge Massachusetts
14. International Committee of Medical Journal Editors (1997) Uniform requirements for manuscripts submitted to biomedical journals. JAMA 277: 927–934
15. Jonsen AR, Siegler M, Winslade WJ (1992) Clinical ethics, 3rd edn. McGraw-Hill
16. Karasz F (1996) Publication ethics and basic sciences. Symposium Proceedings on Publication Ethics. Published by TUBITAK, Ankara
17. Petersdorf RG (1982) Preventing and investigating fraud in research. J Med Educ 57: 880–881
18. Petersdorf RG (1986) The pathogenesis of fraud in medical science. Ann Int Med 107: 252–254
19. Relman AS (1985) Dealing with conflicts of interest. N Engl J Med 313: 749–751
20. Responsible Science (1992) Ensuring the integrity of the research process, vol. I. Prepared and published by the National Academy of Sciences, USA, Washington DC
21. Shapiro MF, Charrow RP (1985) Scientific misconduct in investigational drug trials. NEJM 312: 731–736
22. Swazey JP, Anderson MS, Lewis KS (1993) Ethical problems in academic research. Am Scie 81: 542–553
23. Thompson DF (1993) Understanding financial conflicts of interest. N Engl J Med 329: 573–576
24. Woolf PK (1986) Pressure to publish and fraud in science. Ann Int Med 104: 254–256
25. World Medical Association (2000) Declaration of Helsinki. Ethical principles for medical research involving human subjects. JAMA 283: 3043–3045

Correspondence: Emin Kansu MD, FACP, Hacettepe University, Institute of Oncology, 06100 Sihhiye Ankara, Turkey.

Acta Neurochir (2002) [Suppl] 83: 17–23

Communication in Science

H. Deda and **H. Yakupoglu**

Ankara University, School of Medicine, Department of Neurosurgery, Ankara, Turkey

Summary

Science must have a common language. For centuries, Latin language carried out this job, but the progress in computer technology and internet world through the last 20 years, began to produce a new language with the new century; the computer language.

The information masses, which need data language standardization, are the followings; *Digital libraries and medical education systems, Consumer health informatics, Medical education systems, World Wide Web Applications, Database systems, Medical language processing, Automatic indexing systems, Image processing units, Telemedicine, New Generation Internet (NGI).*

Keywords: Communication in science; computer network; next generation internet; consumer health information; icd10, hl7; telemedicine; Hyper text markup language; world wide web.

Introduction

Science can be science only when it shares its product and uses a common language for this process. In every scientific area, its own communication ways are developed. The Latin language has been a common language in medicine for centuries and has been able to meet the necessities of its era. So till the first half of the 20th century, it was necessary to know Latin in order to study medicine.

With the computer era, a new page has been opened for medicine. Now we live a period of time which we witness the human being trying to create a common science language apart from Latin for communication via the computers. At the beginning of this period there had been great chaos while creating a new language, especially in the last 20 years, but a system began to settle down with the development of the standards.

The simplicity and difficulties of creating a common language for medicine is different from other scientific areas because of its inner dynamics. Medicine is familiar with the standardization. Scales and scores are the signs of the effort to create a common language. On the other side, the variability of the number of the data types cause a different difficulty like breaking the customs. Though they are same at the basement, the way to take patient's history may differ for every physician according to the university or institute he is bound to.

The quality of the researches correlates with the size of the data group. So, the database systems must be compatible with each other and must use a common language. This can be managed by the standardization of data area [1, 2].

The information masses, which need data standardization, are the followings:

Digital Libraries and Medical Education Systems

With the improvement of digital library systems, it is now easy to find the results of previous researches. In this group;

a. Bibliographic information
b. Citations
c. Full-text articles

Can be studied.

Overview of the Digital Library System

- The structure of information and sets of digital objects

The purpose of the information architecture is to represent the riches and variety of library information, using the building blocks of the digital library system.

From a computing view, the digital library is built up from simple components, notably digital objects. A digital object is a way of structuring information in digital form, some of which may be metadata, and includes a unique identifier, called a handle. However, the information in the digital library is far from simple. A single work may have many parts, a complex internal structure, and one or more arbitrary relationships to other works. To represent the complexity of information in the digital library, several digital objects may be grouped together. This is called a set of digital objects. All digital objects have the same basic form, but the structure of a set of digital objects depends upon the information it represents. Information of a digital library can be divided into categories, e.g.: text with SGML mark-up, World Wide Web objects, computer programs, or digitized radio programs. Within each category, rules and conventions describe how to organize the information as sets of digital objects. For example, specific rules will describe how to represent a digitized radio program. For each category, the rules describe the digital objects that are used to represent material in the library, how each is represented, how they are grouped as a set of digital objects, the internal structure of each digital object, the associated metadata, and the conventions for naming the digital objects [3].

Components of the Computer System. The digital library framework permits many different computer systems to coexist. The key components are listed below. They run on a variety of computer systems connected by a computer network, such as the Internet.

User Interfaces

Both the pilot and the prototype have two user interfaces: one for the users of the library, the other for the librarians and system administrators who manage the collections. Each user interface is in two parts. A standard Internet browser is used for the actual interactions with the user. This can be Netscape Navigator, Microsoft's Internet Explorer, or the Grail browser developed by our colleagues at CNRI. The browser connects to client services, which provide intermediary functions between the browser and the other parts of the system. The client services allow the user to decide where to search and what to retrieve; they interpret information structured as digital objects; they negotiate terms and conditions, manage relationships between digital objects, remember the state of the interaction, and convert among the protocols used by the various parts of the system [4].

Repository

Repositories store and manage digital objects and other information. A large digital library may have many repositories of various types, including modern repositories, legacy databases, and Web servers. Section 4 of this report describes the pilot repository that we have implemented and enhancements planned for the prototype. The interface to this repository is called the repository access protocol (RAP). Features of RAP are explicit recognition of rights and permissions that need to be satisfied before a client can access a digital object, support for a very general range of disseminations of digital objects, and an open architecture with well defined interfaces.

Handle System

Handles are general purpose identifiers that can be used to identify Internet resources, such as digital objects, over long periods of time and to manage materials stored in any repository or database. CNRI's handle system is a computer system that provides a distributed directory service for identifiers (handles) for Internet resources. When used with the repository, the handle system receives as input a handle for a digital object and returns the identifier of the repository where the object is stored.

Search System

The design of the digital library system assumes that there will be many indexes and catalogs that can be searched to discover information before retrieving it from a repository. These indexes may be independently managed and support a wide range of protocols. The pilot system is independent of any search system; the prototype is being linked to CIIR's In Query system, which is already in use at the Library of Congress [3].

Outline of the Information Architecture

The Structure of Information in a Digital Library. Interactions, such as the query described above, require that information in a digital library be organized effectively. Within the library, information is stored as basic units of digital information, e.g., a digitized map, a section of text, a Web page, a scanned photograph, etc. In digital form, each basic unit is a sequence of bits, but users often want to refer to material at a higher level of abstraction than the individual item. Common English terms, such as a "report", a "computer program", or an "opera" can refer to many items that are variants of each other. They may have different formats, minor differences of content, different usage restrictions, and so on, but for some purposes users are willing to consider them as equivalent.

The issues to be addressed in structuring information include the following.

Digital materials are frequently related to other materials by relationships such as part/whole, sequence, etc. For example, a digitized text may consist of pages, chapters, front matter, an index, illustrations, and so on. In the World Wide Web, a typical item may include several pages of text, with embedded images, and links to other information. A single computer program is assembled from many files, both source and binary, with complex rules of inclusion. Materials belong to collections. These may be collections in the traditional, custodial sense; they may be the on-line groupings provided by a publisher; or they may be the pages maintained by a Webmaster.

The same item may be stored in several digital formats. Sometimes, these formats are exactly equivalent and it is possible to convert from one to the other (e.g., an uncompressed image and the same image stored with a loss-less compression). At other times, the different formats contain different information (e.g., differing representations of a page of text in SGML and PostScript formats) [5].

Because digital objects are easy to change, different versions are created continually. (Some organizations change their Web home page several times per month.) Versions may differ by a single bit or may be very different. When existing material is converted to digital form, the same physical item may be converted several times. For example, a scanned photograph may have a high-resolution archival version, a medium quality version, and a thumbnail.

Each element of digital information may have different rights and permissions associated with it.

The material may depend upon the characteristics of computer systems and networks, and the size of the material. For example, a user connected to the digital library over a high-speed network may have a different pattern of work from the same user when using a dial-up line.

The information architecture described here provides a general approach to organizing the material within the digital library in such a manner that computer programs can understand the structure of the material and carry out the interactions that the user wishes [6, 7].

Indexing Information

A citation index catalogues the citations that an article makes, linking the articles with the cited works. Citation indices were originally designed mainly for information retrieval and to allow navigating the literature in unique ways, such as backward in time (through the list of cited articles) or forward in time (to find more recent, related articles).

Citation indexing can improve scientific communication by

- revealing relationships between articles,
- drawing attention to important corrections or retractions of published work,
- identifying significant improvements or criticisms of earlier work, and
- helping limit the wasteful duplication of prior research.

Citation indices can also be used to analyze research trends, identify emerging areas of science, and find out where and how often a particular article is cited.

Currently available and proposed citation indices of scientific literature, however, depend heavily on human preparation or editing of information.

Imagine a citation database was freely available over the Internet and was updated every day with all the new works published that day, including papers in traditional and electronic journals, conference papers, theses, technical reports, working papers, and preprints. Such a database would fundamentally change how scholars locate and keep current with the works of others. In turn, this would also affect how scholars publish their own works, in light of the increased visibility of research regardless of publication venue and the increased potential to demonstrate the value of works through citation analysis. In short, a universal citation database would serve as an important catalyst for reform in scholarly communication [8].

Citation Analysis and the Evaluation of Scholar Work. Beyond the benefits for literature research, the analysis of citation data also has an important role in the evaluation of scholarly work and institutional decision making. One study reported that 35% of biochemistry departments and 60% of sociology department's survey had directly used citation data for hiring, promotion or salary decisions. Citation measures have also been used for rankings of entire academic departments; these rankings can influence applicants for graduate studies and major funding initiatives. Citation analysis has even been used to bring senior academic administrators into account by questioning the overall pattern of salary decisions.

Citation analysis also has important indirect effects on scholarly evaluation through citation-based journal rankings. Journal Citation Reports annually ranks journals based on such measures as journal impact factor (the average number of citations received per article published in the journal cited half-life (the number of years covered by the most recent half of a journal's citations) and others. These citation measures are often used by research librarians in making or recommending journal purchasing or cancellation decisions. Scholars are often evaluated by the prestige of the journals in which they publish; both the citation rankings themselves and the availability of journals on library shelves are contributing elements in establishing journal prestige.

However, there are serious methodological issues in the application of citation analysis to scholarly evaluation. Many studies have been criticized on the grounds that citation counts are sensitive to "fads, foibles and popular trends in science" and that simple-minded citation counts are often used without correlation of those counts with other relevant data. Furthermore, usage of the existing citation indexes tends to overemphasize the role of the particular journals indexed and devalue all other forms of scholarly communication.

These inadequacies may well be ameliorated, however, with the development of a freely available universal citation database. The universality of the database would value all forms of publication equally, allowing the impact of works to be judged without measurement bias imposed by the inclusion or non-inclusion in present academic indices. The free availability of the database would allow data to be easily correlated with other information relevant to the evaluation of scholar works.

Consumer Health Informatics

To obtain information about themselves and their illnesses is the most natural right of the people who get health care. The use of modern computers and telecommunications to support consumers in obtaining information, and analyzing their unique health care situations helps them make decisions about their own health.

Categories or types

- *specific information about the user's unique medical situation*
  family history, wellness/risk assessment, behavior
- *allow the user to communicate and interact with health care providers and other users*
  e-mail, discussion groups, support groups
- *provide health information to user*
  encyclopedias, health articles, appointment reminders

For example; "The Consumer Health Information Center (CHIC) is an advice service set up to help you to understand more about health, including hints and tips to stay healthy and ways to get back on your feet when illness strikes.

There are many minor illnesses which, although unpleasant, don't require a visit to the doctor. So who do you ask for speedy advice on what to do and if the bottle you have in the bathroom medicine cabinet is still OK? How do you know if the illness is proceeding normally and if the old wives' tales are really myths or useful tips?

Well, you can now get all this information from the Consumer Health Information Center. The Center gives you access to independent professional advice through an expert panel of doctors, nurses, pharmacists and consumer information experts. The who's who on the panel is available for you to read in the Panel Profile section." [9]

Resources of the Consumer Health Information Service include:

- *an up-to-date collection* of consumer health books, pamphlets, magazines, newsletters and professional journals
- *electronic databases*, such as Health Reference Center and All-Health Watch
- *information files* on specific health topics including selected newspaper clippings, pamphlets, reading lists, popular magazine articles, review articles from professional journals and other materials
- *brief resource guides* on specific health topics listing a selection of books, magazine articles, and self-help groups
- *reviews of consumer health materials* from authoritative sources
- *lists of consumer health books*, magazines and other materials that would be useful to public libraries,

health libraries, health agencies and health care providers across the province

- *other specialized collections* of medical literature when required; and
- *self-help groups*, community and provincial health organizations and agencies and other health-related agencies and associations.

Medical Education Systems

With the usage of communication tools by everybody, in order to protect the healthy people from illnesses, the patients' experiences who had definite illnesses are made reachable by transferring them into a database form. The openings of medical information to common knowledge by professionals and patients' eagerness to interact with each other are important sources for patient education. Now the problem is, to pick up this information from the web [10].

World Wide Web Applications

With www the data about the patients can easily be transferred. For example, with the web page www.skullbase.net the neurosurgeons who deal with this subject can easily contact each other and share their experiences. Being the most visual and known part of Internet, the World Wide Web began to create the basement of a number of applications via its technological improvement. Now, a great number of databases are available on the web and due to its multimedia properties have a place in all medical applications [11].

Database Systems

It is clear that a common language should be used for patient database. Studies have been made to form a common database recording system for 20 years. The greatest difficulty is the great variability of database systems, which must have a common language. Creating ICD-10, HL-7 partly overcame this difficulty. With these two projects, the areas and the size of data, which a database should consist of, are defined [5].

Medical Language Processing

While talking about database systems, we have mentioned that the greatest difficulty is creating a common language. If a database is being created for the first time, it is partly easy to overcome this difficulty. But if a database, which has already been created, has to be updated, then you are in big trouble. You have to overcome both technical problems and personal resistance. Changing of database, which has been used for years, may be a considerable problem for a great number of people. Now by using HTML, which is used at the beginning and SGML, which has largely been standardized recently, medical literature can be classified, and on the way to form a common database great steps have been taken [8, 12].

Automatic Indexing Systems

The standard indexing of the information will allow evaluation in order to standardize research programs. This will fasten the researches in research programs. In the beginning, limited data were studied but now, because the data tends to increase in an uncontrollable manner, indexing has become a necessity. However, after a while this also will cause data accumulation. The improvement of data indexing tools will enable to solve this problem.

Image Processing Units

With the rapid progress in medical imaging tools and almost total settlement of the system into the computer base, transferring the images to digital systems is not very difficult. One of the problems is, there are many programs to standardize visualization of data on windows, and especially Microsoft is working on these specific programs. When a common language is produced, evaluating the data will be much easier, for example the images of Philips and GE will not combat each other. Another problem is transporting of these data because, with their classification and transferring to database, they occupy great areas. But via the specifically developed systems and new data transporting protocols, the problem has largely been solved. But it is still not as common as it should be.

Telemedicine

Telemedicine is the resource to advocate promotion of medical care for consumers and health professionals via telecommunications technology. In the near future with teleconferences both the improvements in research programs and the treatment of patients will be immediately and easily evaluated. The Health Information Society Technology (HIST) study distinguishes 5 types of telemedicine services, which are telepathology, tele-radiology, distant medical education and tele-consultation, tele-emergency services, and tele-surgery. We can name it long distance consultation system. For the time being, the system is managed in its full-meaning in a few centers but will be more common in the future. It will be available to consult on a pathologic preparation or radiological section or physical examination of a patient from a long distance. Surgical manipulations are being made by robotic technology from long distance, as a new technology. Technology will enable to consult and be cured by any physician and medical center all around the world.

New Generation Internet (NGI)

To succeed all of these via the present background of Internet is very difficult because of the big traffic of trading applications. So, an Internet, which uses a different band for publication and compatible only with scientific applications, is needed. Studies about protocols and standards, which will work on NGI, are going on. Without financial support it is impossible to materialize the items above. Technological developers, Internet service providers, health organizations, drug industry can be sponsors for this aim.

Archive systems, improved with big efforts, entail new problems. Which archive systems should be used? Who should reach the archives and in what proportion? So, volunteer people and organizations and professional system developers must work in full support and harmony. In such a way rapid and compatible archive systems, which will meet everyone's necessities can be developed [13, 14].

Issues and Options

Access

- *support schools, libraries, community development*
- *address user needs (elderly, handicapped, portability)*
- *remote areas*

Cost

- *support for research, preventative medicine*
- *partnerships between public and private business*
- *maintenance and upgrades*

Information Quality

- *bias developer or provider*
- *indirect "censorship"*
- *misleading information or outdated information*
- *effects of self-diagnosis and self-treatment*
- *role of development teams*

Security and Privacy

- *"shared" personal health information*

Sharing of information is the most important source for producing new information. Information is shared most rapidly and directly by computer based systems. The greatest disadvantage of it is the difficulty of obtaining the right information. The size of the data pool and the wrong and non-scientific data it unfortunately contains make it more difficult to obtain the right information every passing day. To overcome this problem NGI and a good working database indexing system is needed [15, 16].

Potential Problems

Not all health and research information are correct!

Unfortunately, there is a great proportion of non-scientific or wrong information on the Internet. To clear the system off these is more difficult than is thought to be and usually needs a professional work.

Conclusion

Computers and communication tools belong to today, not to the future. It is not that much easy to foresee the technology of 10 years ahead. We have to accept that time goes by more rapidly since the last century. All through history, human beings didn't have much difficulty in perceiving the next century's inventions and changings. But today, we can make giant-sized computer processes of 20 years ago thousands times more rapid with nail-size microchips. Neuronavigators as large as a notebook and robotic surgical technologies make it difficult for us to foresee how the 20 years ahead will be.

As Kemal Atatürk said in the 1930's

"working does not only mean to get tired for nothing. Technology and science must take advantage of the civilized inventions in order to keep up with the needs of the time"

So to keep up with the needs of the time, communication in science is an important issue.

References

1. Science and Communication: An Author/Editor/User's Perspective on the Transition from Paper to Electronic Publishing. http://www.library.ucsb.edu/istl/98-summer/article3.html.
2. Science and Research Communications. http://www.vanderbilt.edu/News/srcomm/
3. www.nlm.gov
4. Center for Communication in Science, Technology and Management North Carolina State University. http://www.chass.ncsu.edu/ccstm/index.html
5. www.hl7.org
6. EPS Electronic communication in science discussion meeting report: Paris. http://iptsg.epfl.ch/ene/ene_apr96_paris_text.html
7. Informatics and Computer-Mediated Communication in Science. http://www.unc.edu/~ivaisman/info-cmc.htm
8. Communication in Science and Technology – an introduction. http://educate.lib.chalmers.se/cth/pathciv/texts/index2.html
9. Department of Communication Science and Disorders. http://www.hsc.missouri.edu/~shrp/csd/csdhome.html
10. Communication in Science and Technology – an introduction. http://internet.unib.ktu.lt/fizika/TEXTS/communication/communic.htm
11. www.skullbase.net
12. www.icd10.org
13. Center for Next Generation Internet. http://www.ngi.org/
14. The Internet NG Project. http://ing.ctit.utwente.nl/
15. Communication in science. http://www.botany.uwc.ac.za/sciwriting/sciwrite/writing1/sld001.htm
16. Informal Communication in Science: Its Advantages and its Formal Analogues. http://www.psrg.lcs.mit.edu/~bvelez/std-colls/cisi/cisi-386.html

Correspondence: Haluk Deda, M.D., Ankara University, School of Medicine, Department of Neurosurgery, Ibni Sina Hastanesi, 06100 Samanpazari, Ankara, Turkey.

Acta Neurochir (2002) [Suppl] 83: 25–32

Anatomy Based Research in Neurosurgery

A. J. Gonçalves Ferreira

Departments of Neurosurgery and Anatomy, University Hospital de Santa Maria, Av: Egas Moniz, Lisboa, Portugal

Summary

The Anatomy Based Research in Neurosurgery is an important part of Applied Neuroanatomy that mainly concerns neurosurgical procedures and related problems, namely neuroimaging, brain mapping and surgical techniques. This includes a wide variety of research fields such as 2-D and 3-D structure visualization and referencing (stereotaxy), intraoperative imaging cartography and targeting (neuronavigation), as well as microsurgical, functional and endoscopic anatomy. The methods involved in this investigation regard predominantly microsurgical proceedings and morphological variability studies, both on living or post-mortem material (the strict neuroimaging techniques remaining a main tool for radiologists). The present paper provides an overview of the main techniques of neuroanatomical research applied to neurosurgery. Their major indications and requirements are described and discussed. Special attention is paid to some anatomical techniques such as microdissection and serial sectioning as well as 2-D and 3-D image procedures that are explained in detail.

Keywords: Neurosurgery; neuroanatomy; applied neuroanatomy; neuroanatomical research techniques.

Introduction

Anatomy Based Research in Neurosurgery is a major field of anatomical investigation concerning clinical neurosciences, i.e. of Applied Neuroanatomy. It is one of the main areas of modern neuroanatomy, the other being Basic Neuroanatomy. Both have the same object, the study of the nervous system structure, but have distinct methods and goals:

Basic Neuroanatomy involves a wide range of disciplines from macroscopic anatomy to histology and cell biology, including ultrastructural and molecular biology, developmental and chemical neuroanatomy [2, 4, 7, 14, 18, 30, 31, 36, 42, 44, 47, 56]. It is above all experimental, so it uses mainly experimental animal models at cellular and molecular levels. It employs special technologies such as confocal microscopy, micromagnetic resonance, autoradiography, stereology, computed morphometry, "in vivo" brain slices [4, 30, 31, 44, 47, 52, 55].

Applied Neuroanatomy is mostly dedicated to human morphology. It is closely related to neurophysiology, neuroimaging, neurology and neurosurgery, and works predominantly at macro and microscopic levels [3, 5, 6, 8, 9, 10, 12, 16, 17, 33, 34, 40, 43, 48, 49, 51, 59]. Its present developments include anatomical and funcional brain mapping [35, 53, 54], brain warping [54], 3-D statistics and imaging [12, 41, 53, 56], intraoperative brain cartography [37, 41] and microsurgical anatomy [20, 23, 25, 32, 34, 39, 48, 50, 57]. Most of these fields depend on high level mathematics, such as 3D probability and fractal calculus, as well as sophisticated computer graphic applications [13, 41, 52, 54].

The methodology of Applied Neuroanatomy research requires, as any applied science, the fulfilment of some principles and tasks that correspond, in practice, to the main steps that constitute the scientific research itself [11, 60]:

I To identify a good problem
II To design an adequate method to solve it
III To have the necessary material and techniques available

These first three items represent the scientific know-how and may be an object of learning; the following steps are:

IV To master the techniques with skill
V To reach significant results
VI To know how to apply them

These, in turn, represent the research workout and usually improve with practice.

It is important to analyse the first points in respect to Applied Neuroanatomy, in order to define its actual scope:

- What kind of problems does it intend to solve?
- What methods does it employ?
- What material and techniques does it use?

The answers to these questions are the main purpose of this paper.

The Scientific Problems

The first question to be addressed in this regard is:

Are there (still) anatomical problems to solve in neurosurgery?

There is a common impression among many people that anatomical research in general, chiefly gross anatomy but also microanatomy, is in a marked decline. Indeed we have watched in the XX century a substantial decrease of the role played by classical anatomy in medical research (and teaching) along with the proliferation of multiple, specialized and increasingly active anatomical fields. Nevertheless, a growing number of innovative outstanding anatomical papers have appeared in neuroimaging, neurology and neurosurgery publications, as well as in Internet data bases, during the last years [19, 32, 39, 46, 50, 57]. They demonstrate clearly that the old anatomy descriptions do not provide an adequate scientific view of many anatomical structures and that further macro and microscopic neuroanatomical research is needed to answer the actual demands of several fields of clinical neurosciences.

The most frequent topics in applied anatomical research in neurosurgery are presently:

- Problems related to neuroimaging and anatomical mapping
- Problems related to neurosurgical procedures

The first kind of problems involve a great variety of research fields, namely:

1) 2-D and 3-D visualization of anatomical specimen or imaging material (often with multimodal image procedures) [13, 16, 22, 46, 55].
2) 3-D anatomical localization and referencing of brain structures, namely deep seated ones (Stereotactic Anatomy) [1, 41, 53].
3) Intraoperative imaging and targeting (intraoperative MR, neuronavigation) and related problems: intraoperative brain shift, cerebral gyrus identification, special tools manufacturing, etc [15, 29, 37, 45, 56].
4) Anatomical and functional brain mapping (brain cartography) including case to case, sex and age-related (4-D mapping) variations [1, 22, 37, 42, 53, 55, 56].
5) Morphometry and variability studies, since simple cerebral asymmetries related to functional brain dominance until complex 3-dimensional variations. For instance the studies on hippocampal atrophy in Alzheimer disease or in medial temporal sclerosis, founded on the need to know the precise normal volume of this structure [21, 42, 56].

The problems related to specific neurosurgical procedures include the following subjects:

1) Microsurgical anatomy of different regions of the nervous system, now a fertile source of anatomical publications concerning vascular, skull base, brain tumors or spinal cord operations [21, 23, 25, 32, 34, 39, 48, 50, 57, 62].
2) Neuroanatomical basis for functional and stereotactic neurosurgery:
   Functional anatomy of structures involved in the surgical treatment of chronic neurological diseases (Parkinson, epilepsy, pain, spasticity, etc);
   Stereotactic approach routes to different targets within the central nervous system [1, 22, 23, 41, 53].
3) Neuroendoscopic anatomy, whose special knowledge became a main tool for the development of minimally invasive neurosurgery [3, 43].
4) Spine and spinal cord anatomical topography, essential to the progress of spinal instrumentation, owing to the marked individual variations of these structures [46].

The Methods

The anatomy based research in neurosurgery employs two main sorts of methods, according to the research material that is used:

- In vivo anatomical methods
- Post mortem anatomical methods.

The "in vivo" anatomical methods are centered on certain neurosurgical procedures already mentioned, such as stereotactic anatomy or intraoperative brain mapping, as well as on multiple neuroimaging modalities and their most recent technical developments:

1) Helical and Angiographic Computer Tomography (CT)
2) Functional and Dynamic Magnetic Resonance (RM); RM spectroscopy and relaxometry
3) Positron Emission Tomography (PET)
4) Multimedia image integration and 3-D image reconstruction.

The most important post-mortem anatomical methods are based on:

1) Macrodissection
2) Microdissection
3) Anatomical Tomography including serial sectioning and 3-D image reconstruction
4) Digital anatomical imaging procedures
5) Anatomical statistics, namely applied to stereotactic and mapping purposes.

Material and Techniques

The "in vivo" neuroanatomical research is actually dominated by the imaging techniques referred above. It interests mainly the neuroimaging specialists [27]. For this reason it falls out of the scope of this review, which deals essentially with the post-mortem applied neuroanatomical research. This research comprehends two distinct groups of techniques:

- Anatomical techniques
- Imaging techniques and variability studies (applied to the anatomical material).

The first includes specimen preparation and preservation, macro and microdissection or serial sectioning and staining according to the respective method. The second comprises image acquisition, 2-D and 3-D image procedures and variability studies.

Anatomical Techniques

1) Specimen preparation [2]:
   The post-mortem anatomical specimen, namely human, must be kept at low temperature (under 4°Celsius) until the collection. This should be performed during the first 48 hours after death unless a marked anatomical deterioration is already present; if neurochemical or pharmacological studies are to be done (for neurotransmitter ou neuroreceptor tracing) it must be accomplished within the first 5 to 10 hours, which usually requires a special permission.

   Specimen destinated to macro or microdissection often needs to be previously injected, generally by arterial or venous catheterism. This procedure, done in cervical or skull base vessels, must be performed carefully, at low speed and under direct vision, to allow the correct intraluminal filling. Special products have to be used – latex or low viscosity silicon rubber – preferably coloured, in order to reach the aspect, consistency, elasticity and retractility as close as possible to the living ones.

   For injection-corrosion techniques, usually employed for vascular or ventricular casts, it is necessary to use harder injection products, like methylmetacrylate or polyester, and then to submit the specimen to strong acid solutions.

   For microangiography, now seldomly used, the vascular injection must use small molecule coloidal radiopaque suspensions of Barium or Iodine compounds, that solidify in the blood vessels lumen at room temperature.
2) Specimen preservation [2, 61]:
   The preservation includes fixation or freezing, and storing of the specimen.

   The fixation is usually done by immersion or perfusion of a fixation medium. Most often a (10% to 30%) acquous solution of formaldehyde, preferably buffered is used, but many others may be applied, such as ethylglycol, gluteraldehyde or phenic glycerine. If it is necessary to keep some elasticity of the nervous tissue, as for microdissection, a softener medium like Winkler solution must be employed, which delays the hardening of the specimen for days to weeks. This is generally associated with vascular injection of latex or silicon rubber.

   The freezing of the specimen is usually done in a fast (one step or fractionated) way using, for example, isopentane or carboximethyl-cellulose in liquid nitrogen; the blast freezing prevent the crystal formation that would damage the tissues and deteriorate the histological imaging. When this is not important, freezing may be done slowly, in a common freezer, preferably using cryoprotectants like glicerol or dimetylsulfoxide, or by means of embedding solutions like polyvinyl derivates (poly-

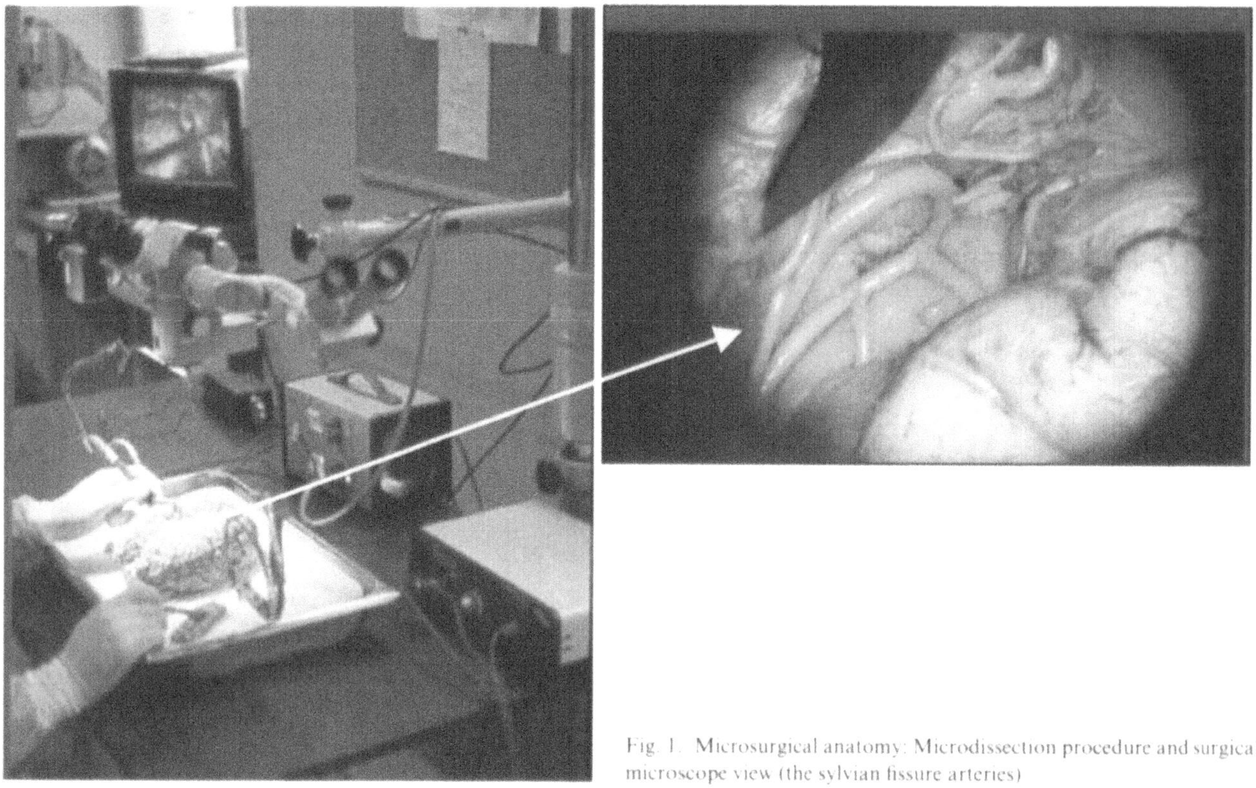

Fig. 1. Microsurgical anatomy: Microdissection procedure and surgical microscope view (the sylvian fissure arteries)

vinylalcohol, polyvinylpyrrolidone) immediately before freezing (see embedding techniques, below).

For special purposes, namely for some teaching preparations, it may be useful to make plastination inclusion blocks of anatomical specimen. This technique consists of embedding the specimen in a transparent plastic solidifying medium that allows a good observation or sectioning with appropriate equipment.

Both chemically fixed, freezed and plastinated preparations need to be adequately stored. Appropriate containers, cabinets, refrigerators and freezers must be available to store the anatomical preparations in good conditions.

3) Specimen dissection (microdissection) [23, 25, 34, 48, 62] (Fig. 1)

The specimen dissection, namely the microdissection, requires first of all a team, which, like a surgical team, consists of at least three people: the principal researcher and two assistants who are in charge of the surgical instruments, the photo and the video devices.

Secondly the microdissection needs a surgical microscope: it does not need to have a zoom objective lens, but it must have an assistant binocular lens and photo/video adapter of the same order of magnification than the main ocular lens; a cold light source is also strongly advisable.

It also requires the appropriate microsurgical instruments: forceps, knife, scissors, dissector, hook, needle handler, all of them being preferably bayonet shaped to minimize the interference with the operating field view; low pression aspirators with thin suction tubes, retractors (self-retaining if possible), vascular clips and clipforceps, bipolar coagulator, microdrill, and a millimetric scale for measurements; other instruments as rongeurs and curettes, as well as sutures, are optional according to the procedures to make.

The photo and/or video coupled cameras are the remaining components of the complete microdissection set. A large depth (and width) of field optic system is preferable as well as, for video and digital photography, an adequate colour balance (never neglecting the different lens magnifications, mentioned above).

Whenever necessary drawing facilities should be available.

Fig. 2. Anatomical and imaging techniques, from serial sectioning to digital imaging: cryomicrotome, freezed embedded block, serial slices mounted and stained, macroscope with coupled digital camera and computer

4) Serial sectioning [2, 19, 22, 26] (Fig. 2)

The nervous tissue sectioning usually requires the previous dissection of the anatomical specimen to get an appropriate anatomical block. Then follows the block positioning for sectioning, sometimes marking it with special fiducials for 3D anatomical referencing (this is generally done in an embedding mould).

The embedding may be done in two different ways: using embedding media that solidify at room temperature, like paraffin or celloidin (both rather retracting when solidifying), gelatin or glycolmetacrylate (these requiring special microtomes), or using embedding products that solidify by freezing such as the carboximethylcellulose or the polyvinyl derivates mentioned above. The latter have the upmost advantage of a physical behaviour until freezing very similar to the nervous tissue, thus leading this to a minimal distortion.

The techniques of nervous tissue serial sectioning most often require a microtome or a cryomicrotome. These should be provided with automatic working programmes and equipped with slice retrieving devices for tape or blade.

After sectioning, the staining is done using the desired techniques: from simple histochemistry to special staining procedures like immune histochemical marking or hybridization.

Then the slice is mounted and observed in the microscope or directly on the computer screen (for digital imaging, see below).

Imaging Techniques and Variability Studies

1) Anatomical image acquisition [55, 56].

The first point to take into consideration in this respect is the image source: microdissection versus serial sectioning anatomical features. In the latter the images are obtained from the cutting surfaces or from the (stained) slices. For both sources attention must be paid to the image magnification, that must always be the same, namely in serial slices, and include millimetric control scales, preferably in two perpendicular directions to prevent the image distortion.

The image acquisitions may be done with optical or digital photo or video equipment (or even a scanner) of high resolution. Anyway it is advisable to use plane apochromatic lens to guarantee the maximal fidelity of the entire anatomical image.

2) 2-D image procedures and variability studies [55, 56].

These image procedures generally include:

1°- Image referencing by anatomical marks or *fiducials* (this is specially important for serial sectioning)

2°- Identification and marking of the structures, namely by tracing their contours

3°- Measurement proceedings: reference marks or *fiducial* coordinates, length(s), perimeter(s) area(s), density(s), cell population(s) etc. (some of these tasks may be performed automatically using special computer programmes)

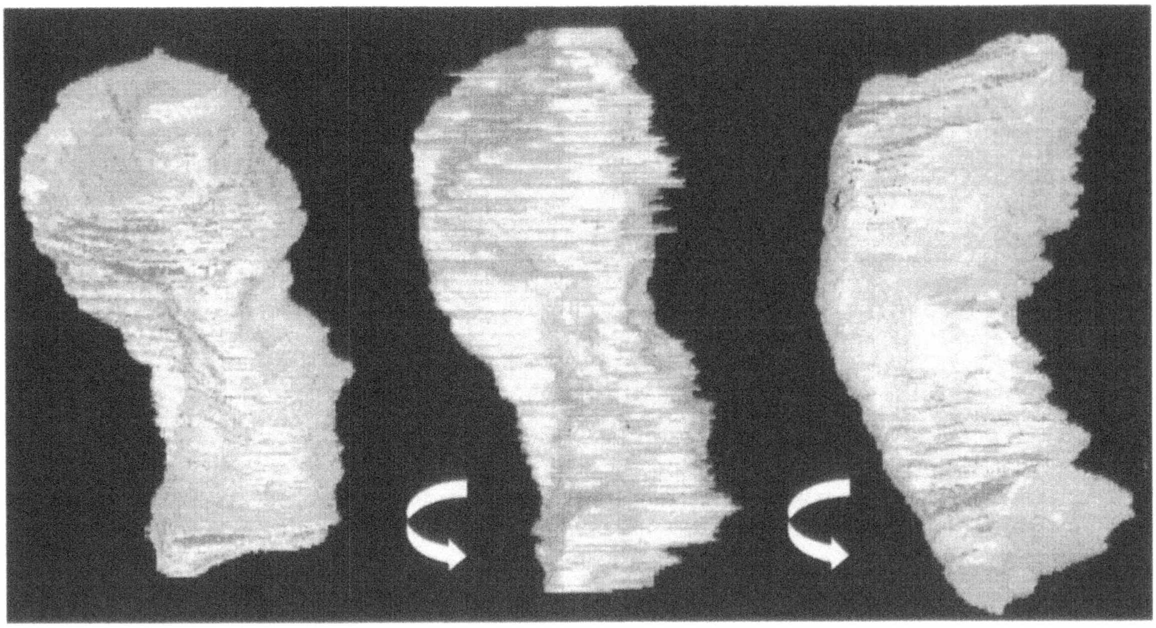

Fig. 3. 3-D reconstruction images of the human hyppocampus at different view angles for a detailed study of its relief

4°- 2-D variability studies, namely referenced anatomical mapping based on 2-D image integration of corresponding cuts of distinct individuals, or elaboration of co-planar stereotactic anatomical atlases.

3) 3-D image procedures and variability studies [37, 54, 56] (Fig. 3)

These usually comprise:

1°- Collection of the serial images with the target structure(s) referenced and marked

2°- Serial images alignment (by reference marks or *fiducials*)

3°- 3-D image structure reconstruction from the contours or from the cutting surfaces of the serial anatomical images, through adequate computer programmes

4°- 3-D variability studies and 3-D probability mapping through:

a) Reconstruction of 3-D images from serial 2-D variability images of all studied individuals;

b) Comparison of 3-D reconstruction images of each single individual obtained with the same 3-D (stereotactic) referencing system.

All these imaging procedures, essentially 3-D anatomical imaging and probability mapping, may be visualized through different sophisticated computer graphic applications such as vector, tensor or probability volume representations, and shown in a wide sort of image editing forms, like multicolour shaded drawings or pictures, or even stereoscopic views.

The main purpose of this paper was to give a global view of the methodology currently used in Anatomy Based Research in Neurosurgery. There was no intention to present a research formulary or an exhaustive up-to-date list of neuroanatomical techniques. The ways to progress in neurosurgery, namely the anatomical pathways, result essentially from repetitive, time consuming, adequately oriented hard work that leads to innovative practice expertise. And this depends essentially on personal capability, team collaboration, research support and, in which concerns neurosurgeons, some degree of surgical activity sacrifice. And still, last but not least, it always relies on a definite, sometimes obstinate, wish to search for further scientific truth.

References

1. Afshar F, Watkins ES, Yap JC (1978) Stereotaxic Atlas of the Human Brainstein and Cerebellar Nuclei. A variability study. Raven Press Books, London
2. Bancroft JD, Stevens A (1996) Theory and practice of histological techniques, 4th edn. Churchil-Livingstone, London
3. Baner BL, Hellwig D (1991) Minimally invasive neurosurgery I. Springer, Verlag
4. Bear MF, Cannors BW, Paradiso MA (2001) Neuroscience. Exploring the brain, 2nd edn. Lippincott Williams & Wilkins, Baltimore

5. Beatty J (1995) Principles of behavioral neuroscience. Brown and Benchmark, Dubuque
6. Bossy J (1990) Anatomie clinique: Neuro-Anatomie. Springer, Wien, New York
7. Brodal A (1981) Neurological anatomy in relation to clinical medicine, 3rd edn. Oxford University Press, Oxford
8. Burst JCM (2000) The practice of neural science. McGraw-Hill, New York
9. Carpenter's Human Neuroanatomy (1996) Williams and Wilkins, Providence
10. Chusid JG (1982) Correlative neuroanatomy and functional neurology, 18th edn. Lange Medical Publications, Los Altos, California
11. Cohadon F (1993) La Recherche en Neurochirurgie. Neurochirurgie 39: 3–6
12. Crosby, Elisabeth C, Humphrey, Tryphena, Leaner, Eduard W (1962) Correlative anatomy of the nervous system. Macmillan Company, New York
13. Damasio H (1995) Human brain anatomy in computerized images. Oxford University Press, New York
14. Delmas A (1991) Voies et Centres Nerveux, 10th edn. Masson et Cie, Paris
15. Dorward NL, Alberti O, Velani B, Gerritsen FA, Harkness WFJ, Kitchen ND, Thomas DGT (1998) Postimaging brain distortion: magnitude, correlates and impact on Neuronavigation. J Neurosurg 88: 656–662
16. Duvernoy HM (1998) The Human hippocampus, 2nd edn. Springer, Berlin Heidelberg New York Tokyo
17. Ellis H (1966) Clinical Anatomy, 3rd edn. Blackwell Scientific publications, Oxford
18. Emson PC (1983) Chemical Neuroanatomy. Raven Press, New York
19. Esiri MM (1996) Oppenheimer's diagnostic neuropathology, 2nd edn. Blackwell Science, London
20. Evans JJ, Hwang YS, Lee JH (2000) Pre-versus post-anterior clinoidectomy measurements of the optic nerve, internal carotid artery and opticocarotid triangle: a cadaveric morphometric study. Neurosurgery 46: 1018–1013
21. Giedd JN, Vaituzis AC, Hamburger SD, Lange N, Rajapakse JC, Kaisen D, Vauss YC, Rapoport JL (1996) Quantitative MRI of the temporal lobe, amigdala and hippocampus in normal human development: 4–18 years. J Comp Neurol 366: 223–230
22. Gonçalves-Ferreira A (1991) Stereotactic anatomy of the posterior cranial fossa; study of the transcerebellar access to the brainstem. Acta Neurochir (Wien) 113: 149–165
23. Gonçalves-Ferreira A, Miguéis J, Farias JP, Levy Melancia J, Andrade M (1994) Selective Amygdalohippocampectomy: which route is the best? An anatomical study on 80 human cerebral hemispheres. Stereotact Funct Neurosurg 63: 182–191
24. Gonçalves-Ferreira A, Farias JP, Herculano Carvalho M, Levy Melancia J, Miguéis J (1995) Corpus callosotomy: some aspects of its microsurgical anatomy. Stereotact Funct Neurosurg 65: 90–96
25. Gonçalves-Ferreira A, Herculano Carvalho M, Levy Melancia J, Farias JP, Gomes L (2001) Corpus callosum: microsurgical anatomy and MRI Surg Radiol Anatomy 23: 409–414
26. Graham DI, Lantos PL (1997) Greenfield's Neuropathology, 6th edn. Arnold, London
27. Greenberg JO (1999) Neuroimaging, 2nd edn. McGraw-Hill, New York
28. Healey JE Jr (1969) Synopsis of clinical anatomy. WB Saunders, Philadelphia
29. Hill DLG, Maurer CR Jr, Maciunas RJ, Barwise JA, Fitzpatrick JM, Wang MY (1998) Measurement of intraoperative brain surface deformation under a craniotomy. Neurosurgery 43: 514–528
30. Kandel ER, Schwartz H, Jessell M (1991) Principles of neural science, 4th edn. McGraw-Hill, New York
31. Kandel ER, Schwartz H, Jessell M (1995) Essentials of neural science and behavior. Appleton & Lange, Englewood Cliff
32. Kim JM, Roman A, Sanan A, van Loveren HR, Keller JT (2000) Microsurgical anatomic features and nomenclature of the paraclinoid region. Neurosurgery 46: 670–682
33. Laget P (1981) Éléments de Neuro-Anatomie Fonctionnelle, 3rd edn. Masson et Cie, Paris
34. Lang J (1980) Clinical anatomy of the head. Springer, Berlin Heidelberg New York Tokyo
35. Maisonnet J, Coudane R (1950) Anatomie Clinique et Opératoire. G Doin et Cie, Paris
36. Marshak DR, Gardner RL, Gottlieb D (2001) Stem cell biology. Cold Spring Harbor, New York
37. Mazziotta SJ (2000) Brain mapping: the disorders. Academic Press, San Diego
38. McMinn RMH (1994) Last's anatomy. Regional and applied, 9th edn. Churchill Livingstone, London
39. Mussi ACM, Rhoton AL Jr (2000) Telovelar approach to the fourth ventricle: microsurgical anatomy. J Neurosurg 92: 812–823
40. Nolte J (1993) The human brain: an introduction to its functional anatomy, 3rd edn. Mosby-Year Book, S Louis
41. Nowincky WL, Thirunavunkarasun A, Kemedy DN (2001) Brain atlas for functional imaging. Clinical and research applications, CD Rom. Georg Thieme Verlag, New York
42. Ono M, Kubik S, Abernathey CD (1990) Atlas of the cerebral sulci. Georg Thieme Verlag, New York
43. Perneczky A, Tschabitscher M, Rescl K (1993) Endoscopic anatomy for neurosurgery. Georg Thieme Verlag, Stuttgart
44. Purves D, Augustine GJ, Fitzpatrick D, Katz C, LaMantia A, McNamara O (1997) Neuroscience. Sinaner Associates, Sunderlan
45. Roberts DW, Hartov A, Kennedy FE, Miga MI, Paulsen KD (1998) Intraoperative brain shift and deformation: a quantitative analysis of cortical displacement in 28 cases. Neurosurgery 43: 749–760
46. Robertson PA, Stewart NR (2000) The radiologic anatomy of the lumbar and lumbosacral pedicles. Spine 25: 709–715
47. Robertson RT (1978) Neuroanatomical research techniques. Academic Press, San Francisco
48. Seeger W (1980) Anatomical dissection for use in neurosurgery. Springer, Berlin Heidelberg New York Tokyo
49. Seeger W (1980) Microsurgery of the brain. Anatomical and technical principles. Springer, Berlin Heidelberg New York Tokyo
50. Seoane E, Tedeschi H, Oliveira E, Wen HT, Rathon AL Jr (2000) The pretemporal transcavernous approach to the interpeduncular and prepontine cisterns: microsurgical anatomy and technique application. Neurosurgery 46: 891–899
51. Snell RS, MD (1997) Clinical neuroanatomy for medical students, 4th edn. Lippincott-Raven, Philadelphia
52. Stewart MG (1992) Quantitative methods in neuroanatomy. John Wiley & Sons, New York
53. Talairach J, Tournoux P (1988) Co-planar stereotaxic atlas of the human brain. Georg Thieme Verlag, New York
54. Toga AW (1999) Brain warping. Academic Press, San Diego
55. Toga AW, Mazziotta JC (1996) Brain mapping: the methods. Academic Press, San Diego
56. Toga AW, Mazziotta JC (2000) Brain mapping: the systems. Academic Press, San Diego

57. Ture V, Yasargil MG, Al-Mefty O, Yasargil DCH (2000) Arteries of the insula. J Neurosurg 92: 676–687
58. Watson C (1977) Basic human neuroanatomy, 2nd edn. Little, Brown and Company, Boston
59. Waxman SG (2000) Correlative neuroanatomy, 24th edn. Langue Medical Books/McGraw-Hill, New York
60. Wilkins RH (1998) Science in neurosurgery: the importance of the scientific method. Neurosurgery 42: 687–691
61. Winkler G (1964) Manual d'Anatomie Topographique et Fonctionnelle. Masson & Cie, Paris
62. Yasargil MG (1984) Microneurosurgery. Georg Thieme Verlag, Stuttgart

Correspondence: A. J. Gonçalves Ferreira, Departments of Neurosurgery and Anatomy, University Hospital de Santa Maria, Av: Egas Moniz, Lisboa, Portugal.

Acta Neurochir (2002) [Suppl] 83: 33–40

Neurological Function-Based Research

A. Savas

Ankara University, School of Medicine, Ankara, Turkey

Summary

Function of the nervous system has been the main domain of interest of an elite group of philosophers, scientists and physicians. This topic may only be understood within a historical perspective, since the progression of knowledge of the function of the nervous system has always paralleled the expansion of our understanding of the outer world and of biological organisms observed. However, data gained from functional experimental and clinical researches in neurosurgery, neurology and neuroscience constitute not only a traditional and historical accumulation of this knowledge but also serve as one of the most important aspects of experimental and clinical neuroscience in the future. This article summarizes the basic principles and methods used in neurological function-based research. In order to elucidate the topic, the text was practically divided into three main sections: (1) historical perspective regarding functional research models in clinical and experimental neuroscience; (2) neurological function-based clinical studies in humans; and (3) experimental research models on animals.

Keywords: Research; neuroscience; neurology; neurosurgery; function; experiment.

Introduction

Researches on the nervous system or observation of the behavior of an organism, as a unit, a part or whole, may help us to understand the rules which relate to the physiology and anatomy of the nervous system. On the other hand, a general understanding of functions of the nervous system is beyond the scope of any single scientific specialty. It still requires philosophical and perhaps metaphysical comments as well, in order to constitute a paradigmatic approach to the concepts of the functioning nervous system. We have always had miniaturized explanations about neuronal functions. However, functions of the neurons are related not only to the basic facts which are open to simple scientific observations, but also to sophisticated concepts such as existence of the individual and outer world. Nevertheless, scientific observation and researches concerning nervous system functions are of great value, being the most reliable data to explain the cognition, behavior and movement of an individual and human perception of the outer world.

Function, biologically, means the special action of an organ, a part or the whole of the body. What represents the function of the nervous system as an action? The answer to this question possibly covers the widest spectrum of the observable characteristics of an organism. Roughly, functional characteristics may be distributed from a membrane of an axon to the behavior of the organism as a whole. Before the beginning of historical information about the progression of scientific knowledge of the functioning nervous system, the following could be stated regarding the acquisition of data, information, hypotheses and ideas about the nervous system:

1. Experimental research is the mainstay, and is accepted as the most reliable way to obtain data regarding the functions of the nervous system and pertinent concepts.
2. Most of the complex functions of the nervous system, such as cognition, language and mood, may be observed better in humans than in the other mammalian species.
3. Findings obtained from neurological patients and from the neurosurgical procedures on humans provide very rich basic neuroscientific information, and lead to an expansion of knowledge regarding the functions of neural structures. On the other hand, some of the methods used in the diagnosis and treatment of patients may be practically applied in various experimental models.

4. Some of the very important conclusions in functional neurosurgery regarding brain functions are gained from empirical or, sometimes, accidental observations.

Historical Perspective Regarding Functional Research Models

Semi-scientific observations or daily-life experiences of humans are common processes directed to understand the functions of the nervous system from the beginning of human existence in the world. Particularly, reflex actions have always been found interesting not only in the sphere of physiology but also in clinical aspect studies. The discoveries in physics and mechanics in the Renaissance lent an intellectual stimulus to attempts to explain the functions of the organisms on mechanical grounds. A group of philosophers and physicians at the time were referred to as "iatrophysicists"; French philosopher Rene Descartes was perhaps one of the most important exponents of these ideas. Descartes wrote a book devoted to the concepts of neurophysiology, and he adumbrated the fundamental ideas in the concept of reflex action. According to Descartes, any stimulus is transmitted along nerve fibers to the brain. There, it gives rise to a new impulse which passes along outgoing nerve fibers to an active organ such as muscle, which is thereby excited to a function. Since then, advances in understanding the function of the nervous system and neurological diseases have been more or less in parallel. Luigi Galvani (1737–1798) was the first experimental researcher who found electricity in an animal and discussed the effects of electricity on nerve and muscles in prepared frogs [8, 31].

German physiologist Johannes Müller (1801–1858) stated the "Law of Specific Nerve Energies" and postulated that each sensory nerve gives rise to its own specific sensation, and any particular form of stimulation produces sensations of light, smell, hearing or taste if applied to the optic, olfactory, auditory, or gustatory nerves respectively. In 1852, Hermann L. F. von (Helmholtz) (1821–1894) measured the velocity of the nerve impulse in the frog (20 m/sec), and later showed that this value is 100 m/sec in humans. Francis Gotch used a capillary electrometer for detection of currents in nerves. The refractory period and all-or-none phenomenon were clearly studied by Keith Lucas. Alan Lloyd Hodgkin investigated nerve transmission using intrafiber nerve microelectrodes (0.5 μ), and showed the basic mechanism of the transmission as a result of an ionic exchange that takes place between the inside and outside of the nerve fiber. He measured the action potential in an axon, and demonstrated that there was resting potential around -70 mV [8, 31].

Russian behavioral physiologist Ivan Pavlov (1849–1936) was investigating the secretion of gastric juice, and he devised an ingenious experimental model to collect the juice at any time after the ingestion of food. He observed that when food was placed in a dog's mouth, the stomach secreted the juice. He concluded that the secretion was controlled, and termed such reflexes unconditioned. He also found that the dog's stomach would secrete juice if the dog was allowed to smell or see food. Here the reflexes were conditioned by the animals. These experiments opened a new gate to evaluate and comment on neuronal activity, which reflects as complex functions of organisms. Sir Charles Scott Sherington (1857–1952), Professor of Physiology at Oxford, reported his experimental studies and found the following: No reflex arc consisted of one single neuron only. There must be not only intracellular but also intercellular conduction of the nerve impulse, thus there must be a barrier between one nerve cell and the next. He postulated that the reflex was the simplest expression of the integrative action of the nervous system, to enable the body as a whole to function to one definite end at a time [8, 31].

Anterolateral cordotomy was first performed in monkeys by Schüller in 1910; application to humans was suggested by Spiller and was developed by Martin in 1911 [14]. Separate innervation of intrafusal and extrafusal fibers of the muscle spindle in mammals was first demonstrated by the ingenious neurosurgeon, Lars Leksell. Leksell used pressure to block conduction in the large motor neurons in ventral roots, so that stimulation of the ventral roots excited only the small diameter motor axons. The excitation produced no significant increase in muscle tension, but did increase the discharge rate of spindle afferents. Thus, he found that another motor system (gamma) modulated the discharge of spindle afferents [12].

The localization of specific functions to different parts of the brain has always been interesting for neuroscientists. After the era of phrenological ideas, which are accepted as a pseudo-science, Marie Jean Pierre Flourens (1794–1831) began to present his experiments on the central nervous system to the Academy of Science in Paris. Flourens gave a classic description of a pigeon from which he had removed the cerebral hemispheres: when only one hemisphere was removed, the bird lost the sight of the eye on the opposite side. He was the first researcher to show in experimental conditions that destruction of a part of the brain may cause a specific function of the nervous system. The French anthropologist and surgeon Pierre-Paul Broca (1824–1880) presented his observations on frontal lobe lesions and speech, and he gave nervous system researchers the theory of cerebellar localization as one of the most important stimuli. Shortly thereafter, English neurologist John Hunglings Jackson (1835–1911) proposed that the nervous system worked as sensory-motor machines and was divided into coordinated centers, following his observations on cases with motor epilepsy and speech disorders. In 1869 H. C. Bastian described the type of impairment of speech known as sensory aphasia, and he localized auditory and visual word centers; five years later the phenomenon was redescribed by Carl Wernicke. Fritsch and Hitzig reported that electrical stimulation of certain points of the cerebral cortex in front of the central sulcus in dogs produced contractions on the opposite side of the body. The human cortex was first stimulated artificially by Robert Bartholow in 1874, and similar human experiments were made by Sir Victor Horsley, the pioneer of experimental stereotactic techniques. Edward Sherrington reported the most careful functional topography of the motor cortex in apes. Caton was the first to record electrical activity from the cortex of cats, rabbits and monkeys. He recorded event-related potentials associated with various stimuli, including odors and tactile and visual stimuli. In 1924, Berger made permanent tracing in human subjects. The human electroencephalogram (EEG) was described for the first time by Hans Berger, a German professor of psychiatry [19, 20, 31]. Canadian neurosurgeon Wilder Penfield worked at the Montreal Neurological Institute and laid the foundation for functional cerebral localization using intraoperative electrocorticography and functional mapping by direct cortical stimulation [27]. In 1935 Fulton and Jacobsen reported their observations of the calming effect of frontal lesions in chimpanzees. Egas Moniz suggested that destruction of the frontal-limbic associations in humans might be a treatment for mental disorders, and he organized a surgical team and performed the first prefrontal lobotomies in humans [17].

Neurotransmitters and transmission of the nerve impulse have been the major route to experimental researches on function of the nerve fibers. The first observations regarding the biochemical events during this transmission were made by French physiologist Claude Bernard (1813–1878), who thought that the transmission across the neuromuscular junction was related to a pharmacological agent rather than purely chemical. While Thomas Elliot (1877–1961) was working on adrenaline, Sir Henry H. Dale (1875–1968) investigated effects of acetylcholine in detail [8, 31]. Neurotransmitter studies directly reflect functional status of the specific neurons and provide data from another aspect.

Clinical Studies in Humans and Function-Based Research

Functional neurosurgery involves a broad spectrum of neurosurgical issues and related techniques directed to investigate neuronal functions. The issues may be summarized as follows: functional procedures for movement disorders, spasticity, surgery to relieve pain or epilepsy, lesioning techniques of neural structures, electrical recording and stimulation techniques, various stereotactic techniques, radiosurgery, etc. Nearly each of these issues is the subject of clinical and experimental research.

As mentioned before, findings obtained from neurological patients and from the neurosurgical procedures on humans provide very rich basic neuroscientific information, and lead to an expansion of knowledge regarding neural structure function. Some of the methods used in the diagnosis and treatment of patients may be practically applied in experimental models.

Very important conclusions in functional neurosurgery regarding brain functions are sometimes gained from empirical or, in some cases, accidental observations. As a classic example, in 1954, Cooper inadvertently tore the anterior choroidal artery on a patient with post-encephalitic Parkinson's disease. Tremor and rigidity were reduced without hemiparesis postoperatively. This observation led to the idea of selective lesioning of deep brain structures [24].

Cerebral functional localization techniques in humans may be summarized as follows:

- Intraoperative cortical mapping:

a) Electrocorticography: specific localization of epileptogenic cortex
b) Cortical electrical stimulation: functional localization of the eloquent cortex

c) Craniotomies in awake patients or under superficial anesthesia: in order to recognize and protect the eloquent cortex

- Evoked potentials:

a) Somatosensory evoked potentials (SEP): to investigate the descending somatosensory tracts, and to recognize the localization of somatic sensory and motor cortex
b) Thalamic-SEP: detection of SEP by direct recording from the thalamus
c) Visual evoked potentials: functional status of the visual pathway
d) Brainstem-auditory evoked potentials: function of brainstem and auditory pathways

- Microrecording, electrical stimulation, and lesioning of the nervous system structures: investigations for the functional localization of the deep brain structures, i.e. thalamus, pallidum, subthalamic nucleus, etc.
- Intracarotid amytal test (WADA): localization of hemispheric dominance
- Electroencephalography: noninvasive investigation of the electrical activity of the cortex from the scalp
- Functional MRI: As blood flow increases in areas of neuronal activation, MRI may detect the decrease in the level of capillary blood deoxyhemoglobin; also gives direct information about the functioning areas of the brain and provides non-invasive functional cortical mapping
- Positron emission tomography (PET): may be used to measure regional cerebral glucose metabolism, blood flow, oxygen consumption and receptor distribution; (^{18}F)fluorodeoxyglucose and (^{15}O)H_2O are the frequently used agents for functional PET mapping
- Magnetoencephalography: Functional areas in the brain and related magnetic fields may be detected with multi-channel magnetometers; combination with MRI systems may give some information about regional activities on the cortex
- Tests directed to investigate specific sensory functions: vision, hearing, etc.
- Some clinical tests and examinations are important in the clinical evaluation of the collective and complex function of the nervous system:

Neurological examination
DSM-III-R code: for evaluation of mental disorders and dementia [1]
Psychological tests; IQ tests
Memory tests
Intracarotid amobarbital test (WADA)
Scales used for movement disorders and disability:

- Hoehn-Yahr scale-parkinsonism
- Unified Parkinson's Disease Rating Scale (UPDRS): parkinsonism (cognition, examination, drug side effects, daily-living score)
- Dystonia movement scale [5]
- Disability scale, etc.
- Scales to measure pain quality: the McGill questionnaire, multidimensional pain evaluation protocol, IASP, etc.

Neurological Function-Based Research on Animals

When we review the methods of researchers working on the functions of neuronal systems, we may notice that a great variety of animals from very different species have been used in experiments. However, it is also remarkable that although researches on the peripheral neurons and the spinal cord could be made on more simple organisms phylogenetically such as amphibians, frogs, etc., complex functions of the central nervous system have usually been assessed in the brains of the mammalians. Rats and rabbits are the animals widely and practically used, but some very important researches have been made on cats, dogs and even anthropoid apes. Work with the higher animals in experimental conditions is becoming more and more difficult due to the public or non-scientific reaction against it.

Basic techniques for research models may be summarized as follows:

- Destructive techniques:

a) mechanical or surgical destruction: cutting a nerve, etc.
b) thermal destruction with radiofrequency or electrocoagulation
c) destruction with radiation (external beam radiosurgery, radionuclide implants, etc.)
d) destruction with chemical agents: phenol, alcohol. Kainic acid has a special role in creating focal destruction in the nervous system [7]
e) Cryogenic lesions

- Electrical stimulation techniques: as mentioned previously

- Various electrophysiological recording techniques: as mentioned previously
- Neuropharmacological manipulation of the neuronal substrates: radiolabeled choline studies to observe the function of the cholinergic neurons [28, 29]
- Transplantation [3, 7]
- Investigations directed to observe a movement as a specific function of a neuronal circuit: walking pattern, pain reaction, etc.
- Investigations of the general behavior of the organism: spatial memory, etc.
- Special techniques used in experiments, e.g. stereotactic frames for animals

Various electrophysiological techniques are used to investigate the function of the peripheral and central nervous system. Electromyography (EMG) and electroneurographies are the basic techniques in clinical and experimental studies, and provide data about the electrical activity of muscles and peripheral nerves. In addition to conventional EMG, motor and nerve conduction studies, somatosensory evoked potentials, reflex studies (H-reflex) and late responses (F-response) may be investigated to observe the functions of the peripheral nerves and central pathways. In most models, animals should be under superficial anesthesia, e.g. fentanyl, which does not suppress electrical activity of the nerves.

We used an experimental model in order to observe the functional improvement of the surgically transected and repaired tibial nerve in rats [32, 33]. Two weeks following tibial nerve transection, rats were re-anesthetized, and a 5 mm portion of nerve containing the neuroma was excised from the proximal stump of the tibial nerve. A graft joining the remaining healthy proximal nerve to the distal stump was made from a portion of the contra-lateral tibial nerve. Functional recovery was assessed by foot-sole sensitivity, walking patterns, and motor nerve and sensory nerve conduction velocities. The sensitivity of the foot-sole of the paw on the operated foot was measured using the method described by De Koning [9], which involves applying a small electrical current of varying intensity (0.2–0.5 mAmp). A normal rat withdraws its paw away from the noxious stimulus as a reflex. Testing was carried out on day 8 to verify that operated rats responded negatively at the maximum stimulus. Starting on day 20, the testing was carried out every other day until rats responded positively to the minimum stimulus on two consecutive days. In another set of experiments, records of the footprints of the rats were obtained by the method of De Medicanelli [10]. A total of three footprints for each rat were measured and the mean values were calculated. The angle values were expressed as percentage of the normal values. Additionally, electrophysiological measurements were carried out under general anesthesia using fluanisone and fentanyl citrate administered subcutaneously. The conduction velocities of the sensory and motor components of the sciatic and tibial nerves were estimated as described by De Koning and Gispen [9].

Spinal reflexes may be assessed in various experimental models in animals, and transection of the spinal cord is a common technique to investigate the function of the spinal reflex circuits. For example, Berkinblit *et al.* transected the spinal cord of frogs, and observed the reflex movements against an irritant to take into account the relative position of their limbs, tracing from film recordings [2]. In another experiment, Pearson *et al.* transected the spinal cord of cats at various levels and investigated rhythmic locomotor patterns [26]. They found that the cats' hindlimbs were still able to function on a treadmill after transection, and that stimulation of the mesencephalic locomotor region was causing a faster gait in decerebrated cats.

Central nervous system destruction may be produced with open surgical or stereotactic techniques. Stereotactic use of radiofrequency systems or injection of kainic acid into the targeted area may be the methods of choice for selective lesioning [7]. Brinkman made unilateral lesions in the supplementary motor area and observed the effects on bimanual coordination in monkeys [4]. Bimanual coordination was tested by the ability of monkeys to push food through a hole with one finger and catch it with the other hand. The lesioned monkey used both index fingers to push the food from top and bottom.

Stereotactic frames were also developed for animals, e.g. for rats and rabbits, and through their utilization, various functional research methods may be employed as representative models of clinical human applications. Stereotactic atlases for the animal brain are widely used for stereotactic localization of brain structures [18, 25].

Animal models may be used in order to research the functional disorders, e.g. Parkinson's disease. The novel neurotoxin, MPTP, is a meperidine analogue sold as a "synthetic heroin" and taken intravenously by drug addicts. The users of this agent manifest the symptoms of parkinsonism, and autopsy findings have revealed a midbrain neuronal loss. 6-hydroxydopamine (6-OHDA), which is given directly into the CSF of rats, and 1-methyl-4-phenyl-1,2,3,6-tetrahydropyridine (MPTP), which is applied systematically to monkeys, are commonly used agents to create artificial parkinsonism in animals. These agents cause progressive degeneration in the nigrostriatal system, and lead to the symptoms of parkinsonism such as resting tremor, rigidity, akinesia and postural instability [6].

Spatial responses and working memory are rather important in the evaluation of high cortical functions in animals. Small lesions around the principal sulcus produce a deficit in the special response task in monkeys, which Harlow *et al.* observed using a system with two containers [17].

Neuronal transplantation aims to restore cell loss in diseases such as parkinsonism and dementia [7]. Björklund showed that fetal mesencephalic dopaminergic grafts were found to survive and partially re-innervate the striatum in rodents and monkeys with artificial parkinsonism [3]. Most of the cholinergic activity in the cortex and the hippocampus is contained in efferents from neurons of the nucleus basalis magnocellularis of Meynert (NBM), located beneath the substantia innominata of the basal forebrain. Autopsy studies in cases of Alzheimer's disease have shown that most of the neurons of the NBM are atropic or decreased in number. Lesions of the NBM in rats reduce the cortical choline acetyltransferase activity by 60–70%. As an experimental model, we made unilateral neurotoxic lesions with kainic acid in the NBM, using a stereotactic system in rats [7]. Ten days after surgery spatial learning and memory were tested by the water maze tasks originally developed by Morris [23]. At the end of the test, the rats with NBM lesions received stereotactic injections of cell suspensions containing tissue dissected from the ventral forebrain of 15-day fetuses into the ipsilateral frontal cortex. For this test, a cylindrical tank (100 cm in diameter, 40 cm deep) was filled to a depth of 30 cm at room temperature, and addition of powdered milk clouded the water. Four starting points 90° apart were marked on the edge of the tank. A transparent glass platform (10 × 10 cm) was placed into a constant quadrant of the tank so that it was 1–2 cm below the water's surface. The pool was located in a corner of a room with many fixed extra maze cues (i.e. bright green flag, window, mirror, etc.). Rats were given two blocks of four trials on each day of four consecutive days. For each trial, the rat was placed in water facing the tank wall at each of the starting points, and was given 120 sec to find the hidden platform and climb onto it. Time to find the platform (escape latency), was measured by chronometer. After 32 trials in four days, the platform was removed. On the 5th day, the rat was placed in the tank at each of the starting points and allowed to swim for 60 sec; the time spent in the quadrant in which the platform was previously placed was measured.

The activity of motor cortical neurons may be detected directly by recording microelectrodes located within the cortex of animals, and this may provide valuable information. Using this technique, Evarts *et al.* found that the activity of motor cortical neurons codes the direction of force loaded [11]. They inserted microrecording electrodes into the motor cortex of a monkey, and set up a mechanical apparatus that permitted the animal to alternatively flex and extend its wrist. EMG recordings were also observed from the flexor and extensor group of muscles. EMGs of flexor and extensor muscles and discharge records of a corticospinal tract neuron were observed under different load conditions. They found absence of neuronal activity with extensor load as an indication of the neuronal coding related with force rather than displacement.

Penicillin is an epileptogenic agent which may be used in functional experiments, when it is applied topically. Musgrave demonstrated in the cat that sectioning the corpus callosum and anterior commissure could abolish penicillin-induced bilaterally synchronous epileptic discharges [21]. Aluminum oxide creams may also be used to induce epileptic discharges in animals; monkeys were used in experimental epilepsy models in the 1950's [15].

Neher and Sakmann developed the patch-clamp technique to record current flow from single ion channels located on the neuronal membranes in frogs [22]. They used a small fire-polished micropipette (1 μm), pressed it against the membrane of a frog skeletal muscle fiber and then measured the current that flows through channels in the membrane under the tip. This has been the major technique used to study the activity and function of various ion channels.

Neurons carry on many metabolic activities. These activities are important for maintaining the functional integrity of the cells. Although these metabolic processes are only indirectly involved in accomplishing the major function of the nervous system, experiments regarding such activities may reflect the functional changes in specific neural cells, even in vitro conditions. Sometimes, identification of these molecules gives very important data related to a specific neuronal function, and manipulations of these neuronal substrates with pharmacological agents may give very specific information about the function of the neurons.

Pharmacological agents may be topically applied to the cortex. For example, Hikosa *et al.* observed in monkeys that the finger coordination of the right hand was severely disorganized following the local injection of muscimol, which is a GABA-agonist and inhibits synaptic transmission, into the somatic sensory cortex. In this experiment, finger coordination was easily detected by testing the monkey's ability to pick up a piece of apple from a funnel [13].

Changes in high-affinity choline uptake in the hippocampus are regarded as an indication of altered cholinergic activity in these regions [16, 30]. The principal advantage of a nervous tissue culture model is that it permits controlled studies regarding continuous observation and direct accessibility of the cells to the pharmacological agents and labeled neurotransmitter precursors. We devised a study to determine the time-dependent course of choline uptake in mature organotypic slice cultures of rabbit hippocampal formation and to assess the effects of continuous and single high-dose irradiation on choline uptake in cultivated slices in vitro [28, 29]. Transverse slices of the hippocampus were dynamically incubated in a cerebrospinal-fluid-like culture medium for 72 hours. To study the changes in choline uptake longitudinally, the slice cultures were processed with 0.1 μM [^{3}H]-choline, and tritium accumulation was counted. In another set of experiments with ^{125}I, 5 μM hemicholinium-3 was used in choline uptake procedures as a competitive high-affinity choline uptake inhibitor. In this experiment, changes in high-affinity choline uptake in the hippocampus served as a very useful functional marker for cholinergic neurons in the hippocampal tissue cultures.

Conclusion

The most prominent tool of clinical neurosurgery or neurology is skill in solving clinical problems. However, this skill must be developed and integrated as a dominant component of neurological research. Experimental and clinical studies on neuronal systems have always attracted an elite group of scientists and physicians. When we review the historical discoveries regarding the functions of the nervous system, we note that such discoveries remarkably affected scientific circles, daily life and opinions of people and, furthermore, the philosophy of their ages.

References

1. American Psychiatric Association (1987) Diagnostic and statistical manual of mental disorders. Washington, DC, p 567
2. Berkinblit MB, Feldman AG, Fukson OI (1986) Adaptability of innate motor patterns and motor control mechanisms. Behav Brain Sci 9: 585–638
3. Björklund A, Stenevi U (1985) Intracerebral neural grafting: a historical perspective: neural grafting in the mammalian CNS. Elsevier, New York, pp 3–14
4. Brinkman C (1984) Supplementary motor area of the monkey's cerebral cortex: short- and long-term deficits after unilateral ablation and the effects of subsequent callosal section. J Neurosci 4: 918–929
5. Burke RE, Fahn S, Marsden D (1985) Validity and reliability of a rating scale for the primary torsion dystonias. Neurology 35: 73–77
6. Burns RS, Chiueh CC, Markey SP *et al* (1983) A primate model of Parkinsonism: selective destruction of dopaminergic neurons in pars compacta of the substantia nigra by MPTP. Proc Natl Acad Sci USA 80: 4546
7. Can MS, Barlas O, Savas A (1993) Amelioration of spatial learning and memory impairment by foetal cholinergic neuronal grafts in rats with lesions of the nucleus basalis magnocellularis. Turkish Neurosurg 3: 6–10
8. Clarke E, O'Malley CD (eds) (1996) Nerve function. The Human brain and spinal cord. Norman Publishing, San Francisco, pp 139–260
9. De Koning P, Gispen WH (1987) Org 2766 improves functional and electrophysiologic aspects of regenerating sciatic nerve in the rat. Peptides 8: 415–422
10. De Medicanelli LL, Freed JW, Wyaatt RJ (1982) An index of the functional condition of rat sciatic nerve based on measurements made from walking tracks. Exp Neurol 77: 634–643
11. Evarts EV (1968) Relation of pyramidal tract activity to force exerted during voluntary movement. J Neurophysiol 31: 14–27
12. Gordon J, Ghez C (1991) Muscle receptors and spinal reflexes. In: Kandel ER, Schwartz JH, Jessel TM (eds) Principles of neural science. Prentice-Hall Int, Inc, pp 564–580
13. Hikosa O, Tanaka M, Sakamoto M, Iwamura Y (1985) Deficits in manipulative behaviors induced by local injections of muscimol in the first somatosensory cortex of the conscious monkey. Brain Res 325: 375–380
14. Kanpolat Y (1998) Percutaneous cervical cordotomy for persistent pain. In: Gildenberg PL, Tasker RR (eds) Textbook of stereotactic and functional neurosurgery. McGraw-Hill, New York, pp 1485–1490
15. Kopeloff N, Kennard M, Pacella B *et al* (1950) Section of corpus callosum in experimental epilepsy in monkey. Arch Neurol Psychiat 63: 719–727
16. Kuhar MJ, Murin LC (1978) Sodium-dependent, high affinity choline uptake. J Neurochem 30: 15–21
17. Kupferman I (1991) Localization of higher cognitive and affective functions: the association cortices. In: Kandel ER, Schwartz JH, Jessel TM (eds) Principles of neural science. Prentice-Hall Int, Inc, pp 823–838
18. Lim RKS, Liu CN, Moffitt RL (eds) (1960) A stereotactic atlas of the dog's brain. Charles C Thomas, Springfield

19. Marino R, Gronich G, Cukiert A (1998) The role of electrocorticography in the evaluation of epilepsy. In: Gildenberg PL, Tasker RR (eds) Textbook of stereotactic and functional neurosurgery. McGraw-Hill, New York, pp 1849–1855
20. Martin del Campo JC, Taub E, Lozano AM (1998) Cerebral localization mapping in neurosurgery. In: Gildenberg PL, Tasker RR (eds) Textbook of stereotactic and functional neurosurgery. McGraw-Hill, New York, pp 1839–1848
21. Musgrave J, Gloor P (1980) The role of the corpus callosum in bilateral interhemispheric synchrony of spike and wave discharge in feline generalized epilepsy. Epilepsia 21: 369–378
22. Neher E, Sakmann B (1976) Single-channel currents recorded from membrane of denervated frog muscle fibers. Nature 260: 799–802
23. Nilsson OG (1990) Intracerebral neuronal transplantation: growth and function of cholinergic neurons transplanted to the hippocampus. University of Lund Press, pp 1–12
24. Parrent AG (1998) Overview of the surgical treatment of movement disorders. In: Gildenberg PL, Tasker RR (eds) Textbook of stereotactic and functional neurosurgery. McGraw-Hill, New York, pp 995–1003
25. Paxinos G, Watson C (1986) The rat brain in stereotactic coordinates. Academic Press, San Diego
26. Pearson K (1976) The control of walking. Sci Am 235: 72–86
27. Penfield W, Roberts L (eds) (1959) Speech and brain mechanisms. Princeton University Press
28. Savas A (1994) The effects of continuous and single dose radiation on choline uptake in organotypic tissue slice cultures of rabbit hippocampus. Doctoral Thesis, Freiburg University, School of Medicine (Freiburg), pp 12–23
29. Savas A, Warnke PC, Ginap T, Feuerstein TJ, Ostertag CB (2001) The effects of continuous and single dose radiation on choline uptake in organotypic tissue slice cultures of rabbit hippocampus. Neurol Res 23: 669–675
30. Simon JR, Kuhar MJ (1976) High affinity choline uptake: ionic and energy requirements. J Neurochem 27: 93–99
31. Singer C, Underwood EA (eds) (1962) The nervous system and neurology. In a short history of medicine. Claredon Press, Oxford, pp 239–280
32. Sporel-Ozakat RE (1990) Peripheral nerve damage and repair: a pharmacological therapeutical approach. Doctoral Thesis, Rijksuniversiteit Utrecht, Utrecht
33. Sporel-Özakat RE, Edwards PM, Hepgül KT, Savas A, Gispen WH (1991) A simple method for reducing autotomy in rats after peripheral nerve lesions. J Neurosci Method 36: 263–265

Correspondence: Doc. Dr. Ali Savas, PK 243 Kavaklidere 06693 Ankara, Turkey.

Acta Neurochir (2002) [Suppl] 83: 41–44

Cellular Molecular Based Research

S. Leenstra

Department of Neurosurgery, Academic Medical Center, University of Amsterdam, Amsterdam, The Netherlands

Summary

In recent years significant progress has been made in identifying genetic alterations in glial brain neoplasms. Nowadays, three types of development to glioblastoma multiforme (the most malignant form of primary brain tumours) can be identified using genetic molecular techniques. Moreover, with these techniques patients can be identified who will respond to the treatment with alkylating cytostatic drugs. Future research on the genome level but in particular on the level of gene expression holds promise for better grading systems, tailored treatment based on genetic profiling and new targets for treatment. In this paper the history of genetic research on glioma and the techniques that are used are briefly reviewed.

Keywords: Glioma; astrocytoma; oligodendroglioma; genetics; pathology; oncogenes; tumour suppressor genes; DNA; loss of heterozygosity.

The spectacular progression in molecular biology techniques is a challenge for researchers in the field of neuro-oncology. To understand the contemporary terminology, sufficient knowledge of the important concepts and techniques in this field is mandatory. Therefore, these issues will be addressed in brief in this paper before discussing the present status of genetic alterations in glial brain tumours.

Oncogenes and Tumour Suppressor Genes

In 1976, DNA sequences almost identical to viral oncogenes were discovered in the genomes of all vertebrate and many invertebrate cells. These genes, named proto-oncogenes or cellular oncogenes, encode growth factors, growth factor receptors, and components of signal transduction mechanisms that mediate the transport of growth factor signals to the nucleus of the cell [3, 28]. The fact that these genes are present in many phylogenetically distinct organisms demonstrates their vital role in cellular growth and development. In neoplastic cells such genes are frequently found to be mutated or overexpressed and are then referred to as oncogenes. In the '80s it became clear from experiments that, at least in vitro, two oncogenes have to be active for malignant transformation [13]. Since then more than 100 different oncogenes have been characterized. A second class of genes related to neoplasia are tumour suppressor genes or anti-oncogenes. Indications for the existence of tumour suppressor genes came from in vitro experiments in which cell hybrids were formed by fusion of neoplastic cells with normal cells. The resulting hybrids lost the malignant features that characterized the neoplastic cells [8]. Based on epidemiological data of retinoblastomas, which occur both in a sporadic and in a familial form, Knudson Jr. (1971) postulated the so-called two-hit hypothesis: two separate genetic events which inactivate the tumour suppressor gene at both chromosomes are needed for tumour formation (Fig. 1) [12]. In the familial form, one mutation (first hit) is already present due to inheritance or a new germ line mutation. The second hit (mutation or loss) occurs somatically in the target tissue (retina). A high penetration of the familial form, which affects both eyes, reflects the high probability of the second hit in the patients. In the sporadic form two somatic events have to occur. Because the somatic mutation rate is low, only one eye is affected in these patients.

The two-hit hypothesis, as formulated by Knudson Jr., was proven by Cavenee *et al.* (1983) who demonstrated loss of the RB1 gene on chromosome 13 in retinoblastoma cells [4]. Thereafter, the retinoblastoma gene was isolated by Friend *et al.* in 1986 [6]. Around 1990, several other tumour suppressor genes were discovered and some of them could also be isolated. One of the best studied tumour supressor genes is the so-called p53 gene, known to play a role (by mutation

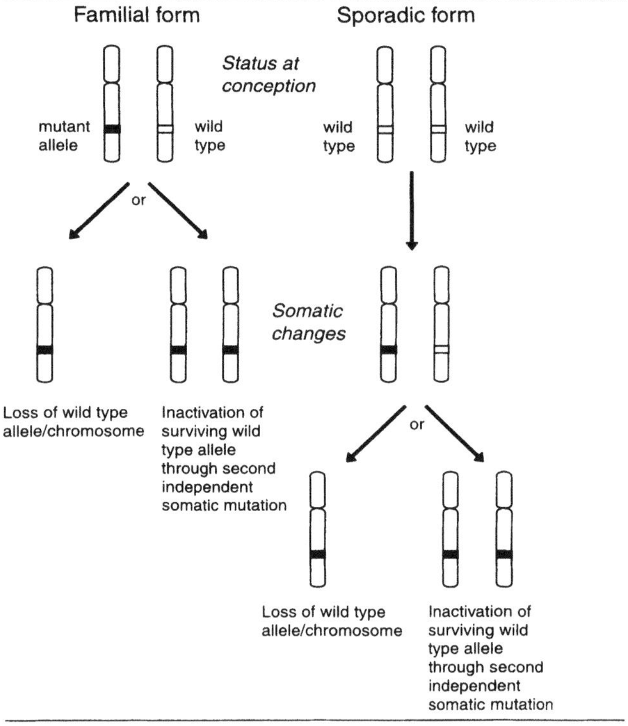

Fig. 1. Two-hit mechanism in sporadic versus familial tumour presentation

and/or deletion) in various malignancies, such as carcinoma of colon, bladder, liver, ovaries, lungs and breast, as well as in astrocytoma [20]. The p53 protein was first detected in rodent cells transformed by SV40 virus. The p53 protein can form a complex with the large T antigen of this virus [26]. Originally the p53 gene was thought to be an oncogene because early experiments with mutant p53 showed transformation of cells in cooperation with the RAS oncogene. Wild type p53 negatively regulates (i.e. inhibits) the cell cycle and the mutant forms of p53 stimulate cell division [17]. Moreover, the p53 wild type gene can reverse the phenotype of transformed cells, glioblastoma multiforme cells included [18]. Experiments with mice embryos and inactivation of the p53 gene by induction of so-called null mutations, have shown a normal gestation but various malignant growths in early life, i.e. sarcomas and lymphomas. Also in human pathology a complex of symptoms is known that is to be related to p53 gene inactivity, the Li-Fraumeni syndrome is characterized by a hereditary p53 mutant gene, as in familial retinoblastoma [16]. Patients with this syndrome may develop carcinomas of colon and breast as well as astrocytoma.

Finally, it is of importance to realise that both oncogenes and tumour suppressor genes cooperate in the process of tumour initiation and progression.

Techniques Used to Detect Genetic Alterations in Glioma

The first studies on genetic alterations in human malignant glioma used cytogenetic analysis [2]. Despite several more or less specific alterations that were found such as loss of chromosome 10 and 22, gain of chromosome 7 and structural abnormalities of chromosome 9p, the results did not integrate in the grading systems for glioma nor did they reveal new strategies for therapy. One of the first techniques to detect involvement of oncogenes or tumour suppressor genes is the technique of Southern blotting [27]. In this procedure, DNA is isolated from the tumour and from normal cells, for instance leukocytes, of the same patient. The DNA is then digested with various restriction enzymes and the resulting DNA fragments are separated by agarose-gel electrophoresis. Thereafter, the DNA fragments are transferred to a nylon filter and hybridized with a panel of DNA markers. By comparing the DNA from normal cells and from tumour cells, amplification of oncogenes and loss of tumour suppressor genes can be detected. In tumour suppressor genes detection depends on the existence of polymorphisms at the loci recognized by DNA markers. If such a polymorphism is present this will result in homologous DNA fragments, originating from the same locus on both chromosomes. The fragments differ in length and, as a result, can be separated by electrophoresis. This principle is known as restriction fragment length polymorphism (RFLP). The homologous chromosomes can be recognized by hybridization with a radioactive DNA marker. If polymorphism is present, i.e. the locus under study is heterozygous for a specific marker, then the differently sized fragments on the two homologous chromosomes are represented as two separate signals on an X-ray film. Chromosomal deletion, including this locus, then results in loss of heterozygosity (LOH) visualized by absence of one of the signals. In the search for tumour suppressor genes, RFLP is often analyzed with a large number of DNA markers covering all 23 chromosome pairs. In some cases, LOH correlates on a specific locus with a specific tumour type or with the malignancy grade. In such a case it is postulated that the locus covers a site that normally harbors a tumour suppressor gene. By using a panel of DNA markers that covers such a region in more detail it is possible to find the site of a tumour suppressor gene.

A major drawback of the two-allele RFLP system is its limitation because, at best, 50% of the patients will be constitutionally heterozygous at a locus. Multiple allele systems can be used, based on the presence of di-, tri- or tetra repeat polymorphisms. The polymorphisms can be visualized by the polymerase chain reaction (PCR) and are far more informative. Nowadays this technique is the standard technique to demonstrate LOH.

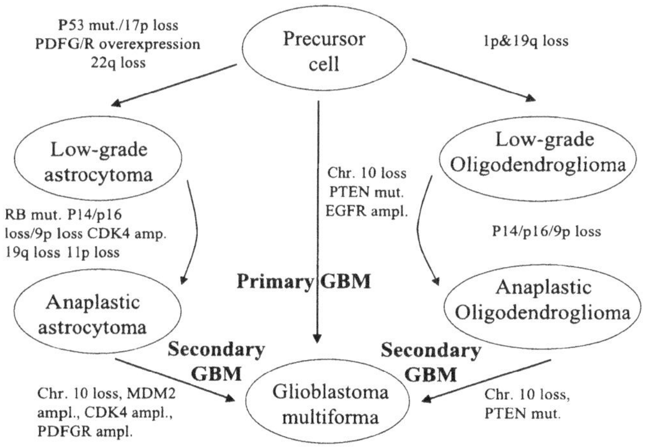

Fig. 2. Three pathways of progression to glioblastoma multiforme show different sets of genetic alterations

Genetic Alterations in Glioma

Among the first genetic alterations that were described were the amplification of the gene coding for the Epidermal Growth Factor Receptor (EGFR), LOH17p and LOH10 [9]. Chromosome 17p harbors the p53 tumour suppressor gene and together with LOH17p, mutations of the p53 gene are frequently encountered in accordance with the two-hit model. In contrast to colon carcinoma, where p53 gene mutation is promptly followed by loss of the remaining allele, 36% of the astrocytomas show LOH17p without mutation of the remaining p53 gene [5]. This further supports the hypothesis of involvement of a second tumour suppressor gene on chromosome 17p. Von Deimling *et al.* found about the same incidence of p53 gene mutations in high grade astrocytomas as in low grade astrocytomas (36 and 37%). This indicates that p53 gene mutations already occur in the initial stage of tumour formation [30]. In addition, Sidransky *et al.* reported a low percentage of cells in low grade astrocytomas with the same mutation as in recurrent high grade tumours, suggesting clonal expansion of these cells [25].

LOH10 is associated with loss of the PTEN [14] and/or DMBT [19] that were both discovered in 1997. Further analysis on other chromosomes has revealed LOH on chromosome 9p [21]. Recent investigations revealed that the tumour supressor genes IKA4a [22] and ARF[1] are involved in the deletions on chromosome 9p.

Finally, an association of LOH with astrocytoma has been reported on chromosomes 11, 13q, 19q and 22 [7]. These genetic alterations are less frequent than LOH10 and LOH17p. Of these alterations, LOH13q and LOH22 are of special interest. The long arm of chromosome 13 harbors the retinoblastoma tumour suppressor gene (RB1 gene), and RB1 gene and transcript alterations were demonstrated in primary glial tumours. Chromosome 22 harbors the neurofibromatosis type II (NF2) gene, a disease with a relatively high prevalence of astrocytomas. Although LOH22 was reported in a few astrocytomas, a recent analysis of 30 astrocytomas (range grade II to grade IV) did not show mutations in the NF2 gene.

To date many alterations in glioma have been found and it became apparent that many of these genes are involved in the cell cycle control (see reference [15] for a review). When the alterations are correlated with clinical data, a model evolves that is depicted in Fig. 2. The model shows three pathways that can result in glioblastoma multiforme, the highest grade of glioma. One form is the so called 'de novo' glioblastoma which is more frequent in the elderly patients with a short clinical history and very short post-operative survival. The other pathways evolve from lower graded tumours, either astroglial or oligodendroglial. These tumour types are more frequently seen in younger patients with a longer history of symptoms and a relatively long post-operative survival. The model is not new, in fact, as early as 1935, Scherer described two distinct forms of glioblastoma multiforme: a primary (arising de novo) and a secondary form (arising from preexistent low grade tumour) [24].

Also of practical importance is that recent investigations show that gliomas having loss of 1p and 19q are more sensitive for alkylating cytostatic drugs such as the combination therapy Procarbazine, CCNU, Vincristine (PCV) or Temozolomide [9]. Future investigations will in addition be directed to gene expression and function of genes. For example, techniques such as Serial analysis of Gene Expression (SAGE [29] and Micro-array analysis [23] can reveal information on over- or underexpression of specific genes in tumours compared to cells from normal tissue. The data of such investigations are now frequently published on the Internet (see for example: http://www.ncbi.nlm.nih.gov/SAGE/). What is the role of the neurosurgeon in these high-tech molecular biology? First of all it is important to set up logistics for tissue sampling, blood sampling (control DNA), storage of these materials and registration of clinical data. Secondly, several members of the neurosurgical community should develop and maintain knowledge of the modern molecular genetic techniques because this enables neurosurgeons to discuss with basic scientists the interpretation of the data and the planning of further investigations. For example, neurosurgeons can point out special cases with an abberant clinical history, casus with specific biological features or provide specific intratumoural sampling guided by neuronavigation, etc. In this type of research it is important that the many data that are generated are correlated with the golden standards such as routine histopathology on the one hand and the clinical history of the patients on the other hand. In this way grading of tumours, for example, can be improved. But even more important: further data on genetic alterations may lead us to mechanisms that might serve as targets for new treatment.

References

1. Arap W, Knudsen E, Sewell DA *et al* (1997) Functional analysis of wild-type and malignant glioma derived CDKN2Abeta alleles: evidence for an RB-independent growth suppressive pathway. Oncogene 15: 2013–2020
2. Bigner SH, Mark J, Bigner DD (1990) Cytogenetics of human brain tumours. Cancer Genet Cytogenet 47: 141–154
3. Bishop JM (1983) Cellular oncogenes and retroviruses. Annu Rev Biochem 52: 301–354
4. Cavenee WK, Dryja TP, Phillips RA *et al* (1983) Expression of recessive alleles by chromosomal mechanisms in retinoblastoma. Nature 305: 779–784
5. Frankel RH, Bayona W, Koslow M, Newcomb EW (1992) p53 mutations in human malignant gliomas: comparison of loss of heterozygosity with mutation frequency. Cancer Res 52: 1427–1433
6. Friend SH, Bernards R, Rogelj S *et al* (1986) A human DNA segment with properties of the gene that predisposes to retinoblastoma and osteosarcoma. Nature 323: 643–646
7. Fults D, Pedone CA, Thomas GA, White R (1990) Allelotype of human malignant astrocytoma. Cancer Res 50: 5784–5789
8. Harris H, Miller OJ, Klein G, Worst P, Tachibana T (1969) Suppression of malignancy by cell fusion. Nature 223: 363–368
9 Ino Y, Betensky RA, Zlatescu MC *et al* (2001) Molecular subtypes of anaplastic oligodendroglioma: implications for patient management at diagnosis. Clin Cancer Res 7: 839–845
10. James CD, Carlbom E, Dumanski JP *et al* (1988) Clonal genomic alterations in glioma malignancy stages. Cancer Res 48: 5546–5551
11. Klinger HP, Shows TB (1983) Suppression of tumourigenicity in somatic cell hybrids. II. Human chromosomes implicated as suppressors of tumourigenicity in hybrids with Chinese hamster ovary cells. J Natl Cancer Inst 71: 559–569
12. Knudson AG Jr (1971) Mutation and cancer: statistical study of retinoblastoma. Proc Natl Acad Sci USA 68: 820–823
13. Land H, Parada LF, Weinberg RA (1983) Cellular oncogenes and multistep carcinogenesis. Science 222: 771–778
14. Li J, Yen C, Liaw D *et al* (1997) PTEN, a putative protein tyrosine phosphatase gene mutated in human brain, breast, and prostate cancer. Science 275: 1943–1947
15. Maher EA, Furnari FB, Bachoo RM *et al* (2001) Malignant glioma: genetics and biology of a grave matter. Genes Dev 15: 1311–1333
16. Malkin D, Li FP, Strong LC *et al* (1990) Germ line p53 mutations in a familial syndrome of breast cancer, sarcomas, and other neoplasms. Science 250: 1233–1238
17. Marx J (1993) How p53 suppresses cell growth. Science 262: 1644–1645
18. Mercer WE, Shields MT, Amin M *et al* (1990) Negative growth regulation in a glioblastoma tumour cell line that conditionally expresses human wild-type p53. Proc Natl Acad Sci USA 87: 6166–6170
19. Mollenhauer J, Wiemann S, Scheurlen W *et al* (1997) DMBT1, a new member of the SRCR superfamily, on chromosome 10q25.3–26.1 is deleted in malignant brain tumours. Nat Genet 17: 32–39
20. Nigro JM, Baker SJ, Preisinger AC *et al* (1989) Mutations in the p53 gene occur in diverse human tumour types. Nature 342: 705–708
21. Olopade OI, Jenkins RB, Ransom DT *et al* (1992) Molecular analysis of deletions of the short arm of chromosome 9 in human gliomas. Cancer Res 52: 2523–2529
22. Quelle DE, Ashmun RA, Hannon GJ *et al* (1995) Cloning and characterization of murine p16INK4a and p15INK4b genes. Oncogene 11: 635–645
23. Schena M, Shalon D, Davis RW, Brown PO (1995) Quantitative monitoring of gene expression patterns with a complementary DNA microarray. Science 270: 467–470
24. Scherer HJ (1935) Gliomstudien. Virch Arch 294: 790–861
25. Sidransky D, Mikkelsen T, Schwechheimer K, Rosenblum ML, Cavanee W, Vogelstein B (1992) Clonal expansion of p53 mutant cells is associated with brain tumour progression. Nature 355: 846–847
26. Soussi T, Caron dF, Mechali M, May P, Kress M (1987) Cloning and characterization of a cDNA from Xenopus laevis coding for a protein homologous to human and murine p53. Oncogene 1: 71–78
27. Southern EM (1992) Detection of specific sequences among DNA fragments separated by gel electrophoresis. 1975. Biotechnology 24: 122–139
28. Stehelin D, Varmus HE, Bishop JM, Vogt PK (1976) DNA related to the transforming gene(s) of avian sarcoma viruses is present in normal avian DNA. Nature 260: 170–173
29. Velculescu VE, Zhang L, Vogelstein B, Kinzler KW (1995) Serial analysis of gene expression. Science 270: 484–487
30. von Deimling A, Eibl RH, Ohgaki H *et al* (1992) p53 mutations are associated with 17p allelic loss in grade II and grade III astrocytoma. Cancer Res 52: 2987–2990

Correspondence: Sieger Leenstra M.D., Ph.D., Department of Neurosurgery, Academic Medical Center, Meibergdreef 9, 1105 AZ Amsterdam, The Netherlands.

Acta Neurochir (2002) [Suppl] 83: 45–48

Basic Research vs. Applied Research

H.-J. Reulen

Department of Neurosurgery, Ludwig Maximilians University, Klinikum Großhadern, Munich, Federal Republic of Germany

Summary

Research rotation is an important component in the education of a neurosurgical resident. The selection of the area of research is essential. There are many arguments in favour of selecting research projects to be close to the individual trainee's clinical interest. Studies far away from the individual's clinical interest in most cases are less productive and will not be pursued later. There are also many advantages if a cooperation is planned with other institutions. The programme director or staff members play an important role in the selection of the research project, of an appropriate laboratory or institution, and in the process of financing a research rotation.

Keywords: Research rotation; categories of research; selection of a research project.

Introduction

In most countries the length of training in neurosurgery is presently 6 years with 4 years obligatory in clinical work and 2 years elective for education in a special neurosurgical area, in neighbouring fields such as neuroradiology, neuropathology, neurosurgical intensive care, or research rotation, respectively. Some programmes already have added a 7th year of training, mostly with the idea to extend the time for clinical neurosurgical education and acquisition of clinical practice. Thus, if we maintain the idea that time should be provided for research in a residency programme, this time has to be used in an optimal manner.

The Goals and Structures of a Research Rotation

In the USA and Germany the majority of the academic programme directors consider a research period as an essential component of education, although most state that not every resident needs a research rotation. If we look at the background of our successful neurosurgical leaders, most – if not all – of them had such an elective exposure to research. The goal of a well organized research rotation is to gain an understanding of the intellectual processes involved in the collection, analysis and interpretation of clinical observations in a systematic way. Such experience will significantly improve the individual's rational approach to the management of patients, the critical judgement of his own work, and also that of reports in the literature (Table 1). Therefore this is an extremely important component of our education [2, 3, 4].

Table 1. *Goals of a Research Rotation*

- Learn to make systematic observations and documentation
- Learn a specific method or technique
- Learn to analyze and interprete the results
- Learn to critically interprete the literature
- Learn to compile the new information in form of a paper
- Become inspired by the excitement of uncovering new knowledge
- Develop independent ongoing research

Different structures are used to organize research rotations. Some departments focus exclusively on research training in a laboratory within that department under the guidance of neurosurgical faculty members who are trained in research. Others prefer to send their trainees to another institution where applied clinical research is sought, particularly if the department does not have appropriate conditions, and finally a third group of departments prefers rotation in an area of basic research in one of the basic neuroscience institutions. Thus, as shown in Table 2, a wide range of possibilities is offered, extending from fundamental basic research in neurosciences to patient oriented studies and finally to randomized controlled trials (Table 2).

Selection of a Research Area

The important question is whether all these options do have the same significance in the context of a neurosurgical training. Can we achieve the above defined goals of research rotation both with a rotation in basic research – for instance on tissue slices – as well as with a participation in a prospective clinical trial.

There are certainly some arguments which strongly support clinically oriented research. This is favoured by most of the experienced programme directors. A successful academic career depends on two columns a) to perform competent clinical patient care and b) to generate high level research. Thus clinical practice and research projects regularly compete for the academic neurosurgeon's time. This is the most challenging competition which can hardly be solved ever by the individual neurosurgeon, particularly if other tasks like administration, teaching, etc. have also to be performed. In contrast to our colleagues in neurology or internal medicine, we spend a large part of the day in the operation theatre. The time remaining for a research project always is limited. Therefore when choosing the research area, there are definite advantages to combine the research with the individual's area of clinical interest, for instance to work on vasospasm if the main clinical interest is vascular neurosurgery. This makes it possible that the individual uses questions posed on a ward round to answer by an adequate study, be it experimental or clinical. The final results of such a study are likely to influence the neurosurgeon's practice.

I strongly believe that a deep change in the understanding and execution of our profession has occurred in the last 25 years. In the 60s and 70s, it was still possible for one person to have an overview of more or less all the technical procedures as well as the theoretical knowledge in our speciality. Studies on the regional blood flow (rCBF) in the vicinity of a brain tumour, on the course of intracranial pressure following severe head injury, on the microanatomy of the branches of the A. cerebri anterior, etc. improved our overall comprehension of the pathophysiological processes or the nature of the disease processes and could be understood by every neurosurgeon. Most of the scientific publications of that time were understandable and of interest for the majority of neurosurgeons.

In the last 10 to 15 years many neurosurgeons tried to increasingly concentrate on specific areas as for instance neurooncology, functional and stereotactic, spinal neurosurgery, skull base surgery, posterior fossa surgery, pediatric neurosurgery, etc., and consequently in all those fields a more profound knowledge was built up involving new technical operative skills and procedures, improvement of the technical tools, etc. Finally this led to the development of subspeciality areas as we have them today in many departments. This is a development which offers patients a higher competence in a specific section, and cannot be reversed any more. In any of those subspecialities the literature has increased considerably so that today it has become impossible to overlook the new information in all subspeciality areas. Furthermore a growing inclination can be observed to plan and organize research projects in a way that they meet the criteria for class I evidence [4].

Table 2. *Categories of Research in Neurosurgery*

A.	*Fundamental basic neurosciences* (blue sky research)
B.	*Laboratory studies simulating clinical diseases* (experimental cerebral vasospasm, brain edema, experimental brain tumors, etc.)
C.	*Applied clinical research* (application of basic research techniques to a clinical problem)
D.	*Research related to technical innovations and to improve surgical techniques* (neuronavigation, functional mapping, laser-technique, instrumentation and biomechanics)
E.	*Clinical observational studies* (retrospective studies, case reports, prospective non-randomized studies, studies on natural course or outcome)
F.	*Randomized prospective controlled trials*

Together with this change in the paradigm, a change in the areas of research took place. There are now much more research areas than we had in the sixties and seventies. In each one of the subspeciality areas there are many possibilities, again ranging from basic research to randomized controlled clinical trials, as shown in Table 2.

As a consequence and in view of the obvious tendency of our young colleagues to focus their clinical work on one of the various subspeciality areas of neurosurgery, the choice of the research topic will become even more important. I will underline this with two examples.

Resident A in his third or fourth year of training, after having received a fairly good overview of most clinical areas, decides that his major interest is spinal neurosurgery, and he, after finishing his regular training programme, wishes to acquire special knowledge in this field by entering a respective fellowship in spinal neurosurgery. It would be logical to choose for his research rotation a project closely related to his future work, in this case for instance spinal neuronavigation, spinal surgical anatomy, spinal biomechanics, instrumentation, etc.

It would be unlogical or even a waste of time, however, if he would be involved for example in studies on hipocampus slices, experimental studies on vasospasm, or any other project far away from his clinical interest. Even with the best introduction to the art of science, the final result of such studies would likely not substantially influence his practice in spinal neurosurgery. It is most probable that at the end of such a research rotation, he would drop these studies. Such a wrong selection of a research project is one of the reasons why many young neurosurgeons after some years of research and successful publication do not pursue their research activities after having received their PhD, a habilitation or the title of a professor. We all know the cases where only over a certain period of time a scientific activity was performed in order to gain an academic title. This is obviously a waste since taking into account the time some individuals invest in research, they would be qualified to instruct younger trainees.

Trainee B at the end of her second or third year of training sees her future commitment in neurooncology and plans to add a neurooncological fellowship after having finished the regular residency programme. After several talks with the programme director she uses her elective time to work 6 months in an oncology department to learn the theoretical and practical principles of chemotherapy and then participates as the local investigator in a multicenter prospective randomised study on brain tumour therapy. There is no doubt that with an appropriate supervision and support this trainee will learn much from such an experience. With this knowledge she will be able to later extend her studies and organize projects by herself. I am convinced that this is one of the important goals, that at a certain point our trainees should become independent and able to organize their own research projects. Such candidates will form the future caders in academic neurosurgery and also form the necessary links to the other neuroscience communities.

The Role of the Programme Director

The consequence of comparing these two examples is that the programme director or an assigned staff member has a high responsibility in the preparation of a research rotation. It is important that he discusses at regular intervals with the trainee his abilities and interests. As soon as the trainee has developed a specific interest for one of the various subspeciality areas, a suitable research topic should be defined. Thus, selection of the research project should be tailored to the individual trainee's clinical interests. The next step would be to choose a well suited laboratory or another department which serves best the needs of the resident for his research rotation. The staff member in such a system certainly plays an important role in the career planning of the resident [3]. Such a procedure will enhance the success of a research rotation and the chances that this trainee will later continue research as part of his career. Such a rational proceeding will no doubt prefer clinically oriented research and reduce purely basic research, as defined previously.

The Dilemma of Clinical Research

This leads to an important problem in our reflections. At this time, financing purely clinically oriented research projects or research rotations by local or national grants is by far more difficult than financing an experimental research project done in cooperation with other institutions. We all have made the same experience. What is the reason for it? Any research project has to be evaluated by several reviewers, and frequently they are representatives of non-clinical disciplines or non-neurosurgical departments. Many interesting projects concerning for example surgical techniques, navigation, instrumentation, etc. or development of new treatment modalities are judged negatively because the reviewer does not understand the underlying clinical questions or cannot be convinced of the quality of the project. This has now become a general problem in most countries. I call it the dilemma of clinical research. On the one side we urgently need good clinical research but on the other side financing becomes more and more difficult.

Is there a way out of this dilemma? One way out for group B and C studies in Table 2 could be a closer cooperation with other research or neuroscience disciplines [1]. This can be realised by either organizing an own research unit run by a basic neuroscientist or seek for collaboration with other neuroscience institutions. Such a cooperation with other disciplines can be fertile in many aspects and for both sides. We can profit from their fully developed and time consuming difficult techniques, we can profit from their knowledge and manpower, and we can apply the techniques and knowledge to our clinical questions. Very often such laboratories or institutions seek for clinical questions and collaboration and therefore a cooperation can serve both purposes. Grant applications for such interdisciplinary cooperations do have a much better chance to receive priority rating. This is particularly valid i.e. for studies relating to tumour biology, tumor immunology, cerebral ischemia, neuro-traumatology and brain pathophysiology. The trend to molecular biology is similar in other clinical disciplines, such as neurology, general surgery, ENT, neurology, etc., and we should consider this when planning the research rotations of our trainees. Generally spoken, it is wise to cooperate closely with one of the basic neuroscience departments, be it anatomy, molecular genetics, immunology, endocrinology, neurophysiology or surgical research, if available. In our department, out of 10 research projects 8 now are done in cooperation with other disciplines. Our growing experience with such cooperations and the experience of many other neurosurgical departments has been very positive and with all probability this is a good chance for the near future.

The expenses for the group of studies as listed under E) in Table 2 are likely to be minimal apart from the trainee's own time, and can be supported by the neurosurgical department. The financing of randomized prospective clinical trials is by far more difficult and can only be organized with the support of pharmaceutical or related companies, respectively, and rarely by national or international grants.

Conclusions

- The research rotation is an important component in the education of a neurosurgical trainee.
- The selection of the area of research should be tailored to the individual trainee's clinical interest.
- Avoid studies far away from the individual's clinical interest.
- Select a proper laboratory, institution or clinical research programme for the research rotation.
- If possible use the advantages of cooperation with other institutions.

References

1. Fahlbusch R (1997) From the scientific idea to its realisation. Principles and strategies in neurosurgery. Acta Neurochir (Wien) [Suppl] 69: 120–125
2. Messmer K, Baethmann A (1997) Role of surgical research in the training of neurosurgeons. Acta Neurochir (Wien) [Suppl] 69: 111–115
3. Reulen HJ (1997) When should residents be exposed to research? Acta Neurochir (Wien) [Suppl] 69: 106–110
4. Teasdale GM (1997) Research in neurosurgical training. Clinical reviews and trials. Acta Neurochir (Wien) [Suppl] 69: 116–119

Correspondence: Prof. H.-J. Reulen, Schwojerstr. 19, D-81249 Munich, Germany.

Acta Neurochir (2002) [Suppl] 83: 49–54

Experimental Models of Head Trauma

R. R. Leker[1], **E. Shohami**[2], and **S. Constantini**[3]

[1] Department of Neurology and the Agnes Ginges Center for Human Neurogenetics, Hebrew University-Hadassah Medical School, Hadassah University Hospital, Jerusalem, Israel
[2] Department of Pharmacology, School of Pharmacy, Hebrew University-Hadassah Medical School, Jerusalem
[3] Department of Pediatric Neurosurgery, Tel-Aviv-Sourasky Medical Center and the Tel Aviv University, Tel Aviv, Israel

Summary

Traumatic brain injury is one of the most common causes for chronic disability in young people. Despite this there are currently no widely available modes of therapy that would limit the extent of brain damage secondary to trauma. Therefore, new insights into the pathological mechanisms involved in head trauma possibly leading to the identification of new therapeutic targets are urgently needed. In order to attain these goals adequate animal models for traumatic brain injury are needed. In the following paper the authors will review the various animal models for head trauma and emphasize their potential strengths and weaknesses.

Keywords: Experimental models; traumatic brain injury; animal.

Introduction

Traumatic brain injury (TBI) is one of the leading causes of mortality and disability in younger individuals [62, 67]. It accounts for an estimated 2 million new cases per year. Much research has been conducted in the field of TBI over the past decades, yet, no specific therapy is available, due to the lack of full understanding of the pathological mechanisms that are involved in the development of secondary brain injury. Numerous treatments are employed today to reduce the neurological damage secondary to TBI, but these are mainly supportive intensive care strategies. Basic research allows for new insights into the pathophysiological mechanisms involved in TBI [24, 48], yet no breakthrough has been made and the pharmacological agents that had been used in clinical trials did not make it into clinical practice thus far. As in other fields of neuroscience (see accompanying review in this issue) most of the basic knowledge was acquired through the use of animal models for TBI [31]. Several such models have been devised over the years with each having its advantages and disadvantages. The purpose of this review is to outline these models.

Different experimental models of traumatic brain injury (TBI) have been devised over the past years [31]. Since TBI is a heterogeneous condition, no single model can depict the actual pathophysiological changes associated with its entire spectrum [47]. Therefore, each model can be seen as representing a subset of injury. Thus, some models are more akin to represent diffuse axonal injury [60] whereas others are more representative of closed head injury with contusions [19] and still others involve traumatic skull fractures with secondary brain impact [11, 51]. In the following pages the authors will try to familiarize the reader with the different experimental models for TBI and to delineate the stronger points of each individual model. Although some in-vitro models for TBI exist (for review see ref [43]) this review is limited to in-vivo models. Using each of these models the interested researcher may evaluate the physiological [2, 18], neurochemical [14, 23, 52, 53, 55], behavioral-cognitive [12, 14, 57], histological [5] and pathological [2, 11, 56] sequela of TBI. Using these methods one can also assess new diagnostic tools and new therapeutic options for neurotrauma [6, 38, 39, 48, 53, 57]. Furthermore, new diagnostic tools such as MRI [3, 4, 7] or MR spectroscopy [10, 59] can be used to further outline TBI pathophysiology.

Table 1. *Characteristics of Experimental TBI Models*

	Species	Technical difficulty	Reproducibility	Severity of TBI	Main pathological features	Applicability to human TBI
Fluid percussion	M/R/C/D/S	medium	high	mild-severe	GM + WM	NA
CHI	M/R	low	high	mild-severe	GM + WM	high
CCI	M/R	medium	high	severe	GM + WM	medium
Rigid indentation	M/R	medium		severe	GM + WM	high
Rotational injury	mini-swine	medium	high	mild-severe	WM	high
Cryogenic injury	M/R/Rab	medium	high	moderate	edema	NA

TBI Traumatic brain injury, *CHI* closed head injury model, *CCI* controlled cortical impact model, *M* mice, *R* rats, *P* primates, *C* cats, *D* dogs, *S* sheep, *Rab* rabbits, *GM* gray matter, *WM* white matter, *NA* not applicable.

The Fluid Percussion Model

This is probably the most widely used model of TBI today. It entails rapid injection of fluids into the cranial cavity [18]. The injected fluids spread along the intact dura mater over the brain and this allows for a rapid rise in intracranial pressure and damage to the brain secondary to its compression [33, 46]. The amount and pressure of the injected fluids and the rate at which they are injected can be controlled to modify the damage [18] and produce mild, moderate or severe injury (Table 1). The pathological correlates of this damage include hemorrhage at the site of injury (e.g. contusion), thalamus, hippocampus and corpus callosum as well as subarachnoid hemorrhage, neuronal and axonal loss and cortical cavitation with reactive gliosis [17]. Another feature of this model includes tissue tears in the white matter [25]. The clinical correlates of this injury are transient apnea or death (depending upon the severity of TBI), supression of postural and non-postural function and locomotor and behavioral abnormalities [18]. These changes are accompanied by hypertension, bradycardia and hyperglycemia, simulating to an extent the changes observed in humans with TBI [18]. The model can be applied to various species allowing for investigation of inter-species variation and evaluation of transgenic animals [8, 9]. Brain injury from fluid percussion may also be combined with other forms of damage such as ischemia or hypoxia allowing for better simulation of the actual conditions observed in humans with TBI.

The advantages of this well standardized model are its high reproducibility and the ability to perform a wide range of severities of brain damage [17, 18]. This model can also be performed in mice allowing for the investigation of knockouts or overexpressors of any given gene [9] and also in other species including cats, piglets, dogs and sheep [40, 46, 64]. Furthermore, the rate of post-traumatic seizures is rather low with this model eliminating another possible source of pathological changes [18].

The disadvantages of this model include the fact that it is invasive and labor intensive. The more severe TBI obtained with this model have a high mortality rate (~50%), and this may lead to an increase in the number of animals needed for each trial [18]. Moreover, although some of the clincal, physiological, pathological and neurochemical changes observed with this model are similar to those observed in human severe head trauma, this model does not replicate the mechanistic features of human TBI.

The Closed Head Injury (CHI) Model

In this model a weight is dropped on the surface of the exposed skull leading to closed head injury [11, 51]. The animals are anathesized and placed on a pad or in a head frame with their skull exposed. A weight is allowed a free fall from a prefixed height, which is determined according to the desired severity of injury. The pathological changes include hemorrhagic necrosis, edema, axonal injury and neuronal loss with secondary gliosis [11, 51]. The pathological scores correlate very well with clinical disability scores and with the degree of brain edema [7, 51]. To minimize the variability in this model, the animals are tested at 1 hour post injury by a set of criteria, collectively called: Neurological Severity Score (NSS). Ten different tasks are used to evaluate motor ability, balancing, and alertness of the mouse, and one point is awarded for failing to perform a particular task. Thus, by including only animals with NSS that reflects the desired severity (severe, moderate or mild) one is able to reach highly reproducible results using this model, which closely mimics the clinical situation. NSS at 1 hour is predictive of both mortality and morbidity and it also correlates with the extent of radiological damage as obtained from T2-weighted MR images (Beni-Adani, unpublished observations). Moreover, in areas identified by MRI as displaying reduced apparent diffusion coefficients (ADCs), a marked reduction in the regional cerebral blood volume (r-CBV) was also observed [4]. These observations may indicate the formation of cytotoxic edema probably due to initiation of ischemic processes. Indeed, reduction in ATP and glucose levels were also noted at the same time points using bioluminescence techniques [36].

Using CHI model several novel therapies were attempted, including dexanabinol [54], which is currently in phase III clinical trial, the endogenous peptide NAP [7] and antisense oligonucleotide (AS-ODN) against acetylcholinesterase (AChE) mRNA that blocked overexpression of the stress-related read-through AChE (AChE-R) mRNA splicing variant after head-injury [53].

The model can be used in rats or mice [11, 12, 63], allowing for investigations of transgenic animals [49, 53, 63]. This model mimics closed (non-penetrating) head injury, and since the location of the injury can be altered (e.g. parietal or frontal) different brain zones can be subjected to injury and the effects of TBI on different areas of the brain can be investigated.

In conclusion, the advantages of this model are its high reproducibility, its being relatively technically non-complicated and easy to implement, and the fact that it may be used in different species. Furthermore, this model closely mimics the physiological and pathological changes observed in human trauma.

The disadvantages of the CHI model are the fact that it may cause, when severe, skull fractures.

Controlled Cortical Impact (CCI) Model

In this model the animals are placed on a foam bed of known spring constant when subjected to an impact [32, 35]. A metal helmet-like device is used to protect from fracture formation. The metal disc usually prevents skull fracturing and also allows for an even distribution of the pressure-impact onto the brain below. The weight and height can be adjusted so as to cause mild, moderate or severe injury with and without edema or contusions [2]. The injury results in motor and behavioral disabilities [1] for which the pathological correlates are those of intra-parenchymatous and subarachnoid hemorrhages, edema and gray and white matter damage [15, 19, 30]. Use of these models usually results in lasting motor and cognitive-behavioral abnormalities [16, 20–22].

These models have been described in various species including rats and mice [15, 61].

This model has the advantages of causing a severe TBI with no accompanying fracture or focal contusions. However, it usually causes a severe injury with a high mortality rate.

The advantages of this model include high reproducibility, ability to cause different severities of injury and ability to investigate lesions in various brain regions. It is also minimally invasive and can be tested in both rats and mice. However, it does not really imitate the mechanics of injury typically seen in human TBI.

Rigid Indentation Models

These models use a rigid impactor that injures the dura and the brain. The pendulum striker model is one of the variants of this model in which a pendulum-striker is used to hit the skull [41]. This leads to neuronal loss without contusions or skull fractures [41]. The degree of indentation and the location of injury can be modified so as to cause mild, moderate or severe TBI at different brain regions. Most currently used modifications of this technique apply a pneumatically driven impactor.

This model has the advantages of being reproducible and relatively easy to perform. It can be used in different species and therefore allow for investigations of TBI in transgenic animals. However, it has the disadvantages of leading to non-specific injury which is usually severe and is associated with high mortality.

The dynamic cortical deformation model is another variant of this model that was only recently described [56]. In this model rats are exposed to transient non-ablative vacuum pulse that lasts 25 msec [56]. This results in a rapid deformation of the cortex and the formation of a cerebral contusion. The pathological changes observed in this model are similar to those observed with human cerebral contusions and include neuronal loss at the site of injury with secondary changes in the subcortical white matter and diffuse astrogliosis of the involved hemisphere [56]. No fractures or axonal injury to the non-involved hemisphere could be discerned [56]. The severity of injury can be determined by the amount of pressure applied. Thus, a pressure wave of 4 PSI leads to moderate injury and a pressure pulse of 8 PSI to severe injury [56].

Because this model is relatively new and has not been widely used, it is difficult to comment at this stage about its advantages in terms of reproducibility. However, it may well be used as a specific model for brain contusion in order to evaluate new therapeutic interventions targeting this form of injury.

Rotational Injury Model

This model was originally developed in the miniswine [37, 60]. Following anesthesia the animals are exposed to impulsive centroidal rotation of 110 degrees in 4–6 msec [60]. This leads to widespread diffuse axonal injury in combination with gliosis that usually affects the interface between gray and white matter and the bases of gyri [60]. This model also leads to minor neuronal damage that is mostly confined to the hippocampus [60]. Notably, no bone fractures, hemorrhage or focal contusions are noted. Since these pathological changes are similar to those observed in humans, this model is used for the investigation of diffuse axonal injury (DAI).

The advantages of this model are its high reproducibility and the fact that it leads to relatively selective DAI without neuronal injury allowing for specific investigation of this condition. It is relatively simple and easily applicable, although it needs a rotational device.

On the down side it is more expensive than other models and its use has been limited to the miniswine. Therefore, investigations using genetic manipulations cannot be performed with this model.

The Cryogenic Injury Model

This model entails the application of cold fluids [28, 29] or metal rods [65] to the exposed brain surface. This leads to the formation of cold injury with early and delayed sequelae [28, 65, 68] the most important of which is brain edema. The model can be used in different species including rats [28, 65], rabbits [68] and mice [42, 44]. It has been used to evaluate neurochemical, metabolic and blood flow changes [13, 27, 29, 34, 44, 45, 50, 58, 66] after lesioning the brain and also to evaluate therapeutic modalities targeting brain edema [26, 68, 69]. However, this model is not widely used since it does not mimic human TBI.

In summary, we provided an overview of the different experimental in-vivo models that are designed to investigate the pathophysiology of traumatic brain injury and to offer therapy to its victims. There are many compounds that appear to be beneficial in experimental studies, yet, in clinical trials were disappointing. Animal models mimic only in part the clinical situations and they should be refined to better represent the human condition. Moreover, pharmaceutical companies that design clinical trials should target the type of brain trauma to the specific model that best represents it. Thus, we believe the rate of success of novel compounds, strategies or therapeutic modalities in clinical trials could be enhanced.

References

1. Adelson PD, Dixon CE, Robichaud P, Kochanek PM (1997) Motor and cognitive functional deficits following diffuse traumatic brain injury in the immature rat. J Neurotrauma 14: 99–108
2. Adelson PD, Robichaud P, Hamilton RL, Kochanek PM (1996) A model of diffuse traumatic brain injury in the immature rat. J Neurosurg 85: 877–884
3. Albensi BC, Knoblach SM, Chew BG, O'Reilly MP, Faden AI, Pekar JJ (2000) Diffusion and high resolution MRI of traumatic brain injury in rats: time course and correlation with histology. Exp Neurol 162: 61–72
4. Assaf Y, Holokovsky A, Berman E, Shapira Y, Shohami E, Cohen Y (1999) Diffusion and perfusion magnetic resonance imaging following closed head injury in rats. J Neurotrauma 16: 1165–1176
5. Baskaya MK, Dogan A, Temiz C, Dempsey RJ (2000) Application of 2,3,5-triphenyltetrazolium chloride staining to evaluate injury volume after controlled cortical impact brain injury: role of brain edema in evolution of injury volume. J Neurotrauma 17: 93–99
6. Belayev L, Alonso OF, Huh PW, Zhao W, Busto R, Ginsberg MD (1999) Posttreatment with high-dose albumin reduces histopathological damage and improves neurological deficit following fluid percussion brain injury in rats. J Neurotrauma 16: 445–453
7. Beni Adani L, Gozes I, Cohen Y, Assaf Y, Steingart RA, Brenneman DE, Eizenberg O, Trembolver V, Shohami E (2001) A peptide derived from activity-dependent neuroprotective protein (ADNP) ameliorates injury response in closed head injury in mice. J Pharmacol Exp Ther 296: 57–63
8. Carbonell WS, Grady MS (1999) Regional and temporal characterization of neuronal, glial, and axonal response after traumatic brain injury in the mouse. Acta Neuropathol (Berl) 98: 396–406
9. Carbonell WS, Maris DO, McCall T, Grady MS (1998) Adaptation of the fluid percussion injury model to the mouse. J Neurotrauma 15: 217–229
10. Cecil KM, Lenkinski RE, Meaney DF, McIntosh TK, Smith DH (1998) High-field proton magnetic resonance spectroscopy of a swine model for axonal injury. J Neurochem 70: 2038–2044
11. Chen Y, Constantini S, Trembovler V, Weinstock M, Shohami E (1996) An experimental model of closed head injury in mice: pathophysiology, histopathology, and cognitive deficits. J Neurotrauma 13: 557–568
12. Chen Y, Lomnitski L, Michaelson DM, Shohami E (1997) Motor and cognitive deficits in apolipoprotein E-deficient mice after closed head injury. Neuroscience 80: 1255–1262

13. Darby JM, Nemoto EM, Yonas H, Yao L, Melick JA, Boston JR (1993) Local cerebral blood flow measured by xenon-enhanced CT during cryogenic brain edema and intracranial hypertension in monkeys. J Cereb Blood Flow Metab 13: 763–772
14. Dietrich WD, Alonso O, Busto R, Ginsberg MD (1994) Widespread metabolic depression and reduced somatosensory circuit activation following traumatic brain injury in rats. J Neurotrauma 11: 629–640
15. Dixon CE, Clifton GL, Lighthall JW, Yaghmai AA, Hayes RL (1991) A controlled cortical impact model of traumatic brain injury in the rat. J Neurosci Method 39: 253–262
16. Dixon CE, Kochanek PM, Yan HQ, Schiding JK, Griffith RG, Baum E, Marion DW, DeKosky ST (1999) One-year study of spatial memory performance, brain morphology, and cholinergic markers after moderate controlled cortical impact in rats. J Neurotrauma 16: 109–122
17. Dixon CE, Lighthall JW, Anderson TE (1988) Physiologic, histopathologic, and cineradiographic characterization of a new fluid-percussion model of experimental brain injury in the rat. J Neurotrauma 5: 91–104
18. Dixon CE, Lyeth BG, Povlishock JT, Findling RL, Hamm RJ, Marmarou A, Young HF, Hayes RL (1987) A fluid percussion model of experimental brain injury in the rat. J Neurosurg 67: 110–119
19. Foda MA, Marmarou A (1994) A new model of diffuse brain injury in rats. Part II: morphological characterization. J Neurosurg 80: 301–313
20. Fox GB, Faden AI (1998) Traumatic brain injury causes delayed motor and cognitive impairment in a mutant mouse strain known to exhibit delayed Wallerian degeneration. J Neurosci Res 53: 718–727
21. Fox GB, Fan L, LeVasseur RA, Faden AI (1998) Effect of traumatic brain injury on mouse spatial and nonspatial learning in the Barnes circular maze. J Neurotrauma 15: 1037–1046
22. Fox GB, Fan L, Levasseur RA, Faden AI (1998) Sustained sensory/motor and cognitive deficits with neuronal apoptosis following controlled cortical impact brain injury in the mouse. J Neurotrauma 15: 599–614
23. Genis L, Chen Y, Shohami E, Michaelson DM (2000) Tau hyperphosphorylation in apolipoprotein E-deficient and control mice after closed head injury. J Neurosci Res 60: 559–564
24. Graham DI, McIntosh TK, Maxwell WL, Nicoll JA (2000) Recent advances in neurotrauma. J Neuropathol Exp Neurol 59: 641–651
25. Graham DI, Raghupathi R, Saatman KE, Meaney D, McIntosh TK (2000) Tissue tears in the white matter after lateral fluid percussion brain injury in the rat: relevance to human brain injury. Acta Neuropathol (Berl) 99: 117–124
26. Hartl R, Schurer L, Goetz C, Berger S, Rohrich F, Baethmann A (1995) The effect of hypertonic fluid resuscitation on brain edema in rabbits subjected to brain injury and hemorrhagic shock. Shock 3: 274–279
27. Hermann DM, Mies G, Hossmann KA (1998) Effects of a traumatic neocortical lesion on cerebral metabolism and gene expression of rats. Neuroreport 9: 1917–1921
28. Husz T, Joo F, Antal A, Toldi J (1992) Late consequences of cryogenic brain lesion in rat; an electrophysiological study. Neuroreport 3: 51–54
29. James HE, Schneider S (1990) Cryogenic brain oedema: loss of cerebrovascular autoregulation as a cause of intracranial hypertension. Implications for treatment. Acta Neurochir (Wien) [Suppl] 5179–5181
30. Kochanek PM, Marion DW, Zhang W, Schiding JK, White M, Palmer AM, Clark RS, O'Malley ME, Styren SD, Ho C *et al* (1995) Severe controlled cortical impact in rats: assessment of cerebral edema, blood flow, and contusion volume. J Neurotrauma 12: 1015–1025
31. Laurer HL, McIntosh TK (1999) Experimental models of brain trauma. Curr Opin Neurol 12: 715–721
32. Lighthall JW (1988) Controlled cortical impact: a new experimental brain injury model. J Neurotrauma 5: 1–15
33. Lindgren S, Rinder L (1969) Production and distribution of intracranial and intraspinal pressure changes at sudden extradural fluid volume input in rabbits. Acta Physiol Scand 76: 340–351
34. Maeda M, Akai F, Yanagihara T (1997) Neuronal integrity and astrocytic reaction in cold injury: an immunohistochemical investigation. Acta Neuropathol (Berl) 94: 116–123
35. Marmarou A, Foda MA, van den Brink W, Campbell J, Kita H, Demetriadou K (1994) A new model of diffuse brain injury in rats. Part I: Pathophysiology and biomechanics. J Neurosurg 80: 291–300
36. Mautes AE, Thome D, Steudel WI, Nacimiento AC, Yang Y, Shohami E (2001) Changes in regional energy metabolism after closed head injury in the rat. J Mol Neurosci 16: 33–39
37. McGowan JC, McCormack TM, Grossman RI, Mendonca R, Chen XH, Berlin JA, Meaney DF, Xu BN, Cecil KM, McIntosh TK, Smith DH (1999) Diffuse axonal pathology detected with magnetization transfer imaging following brain injury in the pig. Magn Reson Med 41: 727–733
38. McIntosh TK, Juhler M, Wieloch T (1998) Novel pharmacologic strategies in the treatment of experimental traumatic brain injury. J Neurotrauma 15: 731–769
39. McIntosh TK, Saatman KE, Raghupathi R, Graham DI, Smith DH, Lee VM, Trojanowski JQ (1998) The Dorothy Russell Memorial Lecture. The molecular and cellular sequelae of experimental traumatic brain injury: pathogenetic mechanisms. Neuropathol Appl Neurobiol 24: 251–267
40. Millen JE, Glauser FL, Fairman RP (1985) A comparison of physiological responses to percussive brain trauma in dogs and sheep. J Neurosurg 62: 587–591
41. Morehead M, Bartus RT, Dean RL, Miotke JA, Murphy S, Sall J, Goldman H (1994) Histopathologic consequences of moderate concussion in an animal model: correlations with duration of unconsciousness. J Neurotrauma 11: 657–667
42. Morita Fujimura Y, Fujimura M, Kawase M, Chan PH (1999) Early decrease in apurinic/apyrimidinic endonuclease is followed by DNA fragmentation after cold injury-induced brain trauma in mice. Neuroscience 93: 1465–1473
43. Morrison B 3rd, Saatman KE, Meaney DF, McIntosh TK (1998) In vitro central nervous system models of mechanically induced trauma: a review. J Neurotrauma 15: 911–928
44. Murakami K, Kondo T, Yang G, Chen SF, Morita Fujimura Y, Chan PH (1999) Cold injury in mice: a model to study mechanisms of brain edema and neuronal apoptosis. Prog Neurobiol 57: 289–299
45. Nag S (1996) Cold-injury of the cerebral cortex: immunolocalization of cellular proteins and blood-brain barrier permeability studies. J Neuropathol Exp Neurol 55: 880–888
46. Pfenninger EG, Reith A, Breitig D, Grunert A, Ahnefeld FW (1989) Early changes of intracranial pressure, perfusion pressure, and blood flow after acute head injury. Part 1: an experimental study of the underlying pathophysiology. J Neurosurg 70: 774–779
47. Povlishock JT, Hayes RL, Michel ME, McIntosh TK (1994) Workshop on animal models of traumatic brain injury. J Neurotrauma 11: 723–732
48. Raghupathi R, McIntosh TK (1998) Pharmacotherapy for traumatic brain injury: a review. Proc West Pharmacol Soc 41: 241–246

49. Sabo T, Lomnitski L, Nyska A, Beni S, Maronpot RR, Shohami E, Roses AD, Michaelson DM (2000) Susceptibility of transgenic mice expressing human apolipoprotein E to closed head injury: the allele E3 is neuroprotective whereas E4 increases fatalities. Neuroscience 101: 879–884
50. Schneider GH, Hennig S, Lanksch WR, Unterberg A (1994) Dynamics of posttraumatic brain swelling following a cryogenic injury in rats. Acta Neurochir (Wien) [Suppl] 60437–60439
51. Shapira Y, Shohami E, Sidi A, Soffer D, Freeman S, Cotev S (1988) Experimental closed head injury in rats: mechanical, pathophysiologic, and neurologic properties. Crit Care Med 16: 258–265
52. Shohami E, Gallily R, Mechoulam R, Bass R, Ben-Hur T (1997) Cytokine production in the brain following closed head injury: dexanabinol (HU-211) is a novel TNF-alpha inhibitor and an effective neuroprotectant. J Neuroimmunol 72: 169–177
53. Shohami E, Kaufer D, Chen Y, Seidman S, Cohen O, Ginzberg D, Melamed-Book N, Yirmiya R, Soreq H (2000) Antisense prevention of neuronal damages following head injury in mice. J Mol Med 78: 228–236
54. Shohami E, Novikov M, Bass R (1995) Long-term effect of HU-211, a novel non-competitive NMDA antagonist, on motor and memory functions after closed head injury in the rat. Brain Res 674: 55–62
55. Shohami E, Shapira Y, Cotev S (1988) Experimental closed head injury in rats: prostaglandin production in a noninjured zone. Neurosurgery 22: 859–863
56. Shreiber DI, Bain AC, Ross DT, Smith DH, Gennarelli TA, McIntosh TK, Meaney DF (1999) Experimental investigation of cerebral contusion: histopathological and immunohistochemical evaluation of dynamic cortical deformation. J Neuropathol Exp Neurol 58: 153–164
57. Sinson G, Perri BR, Trojanowski JQ, Flamm ES, McIntosh TK (1997) Improvement of cognitive deficits and decreased cholinergic neuronal cell loss and apoptotic cell death following neurotrophin infusion after experimental traumatic brain injury. J Neurosurg 86: 511–518
58. Siren AL, Knerlich F, Schilling L, Kamrowski Kruck H, Hahn A, Ehrenreich H (2000) Differential glial and vascular expression of endothelins and their receptors in rat brain after neurotrauma. Neurochem Res 25: 957–969
59. Smith DH, Cecil KM, Meaney DF, Chen XH, McIntosh TK, Gennarelli TA, Lenkinski RE (1998) Magnetic resonance spectroscopy of diffuse brain trauma in the pig. J Neurotrauma 15: 665–674
60. Smith DH, Chen XH, Xu BN, McIntosh TK, Gennarelli TA, Meaney DF (1997) Characterization of diffuse axonal pathology and selective hippocampal damage following inertial brain trauma in the pig. J Neuropathol Exp Neurol 56: 822–834
61. Smith DH, Soares HD, Pierce JS, Perlman KG, Saatman KE, Meaney DF, Dixon CE, McIntosh TK (1995) A model of parasagittal controlled cortical impact in the mouse: cognitive and histopathologic effects. J Neurotrauma 12: 169–178
62. Sosin DM, Sniezek JE, Waxweiler RJ (1995) Trends in death associated with traumatic brain injury, 1979 through 1992. Success and failure. Jama 273: 1778-1180
63. Stahel PF, Shohami E, Younis FM, Kariya K, Otto VI, Lenzlinger PM, Grosjean MB, Eugster HP, Trentz O, Kossmann T, Morganti-Kossmann MC (2000) Experimental closed head injury: analysis of neurological outcome, blood-brain barrier dysfunction, intracranial neutrophil infiltration, and neuronal cell death in mice deficient in genes for pro-inflammatory cytokines. J Cereb Blood Flow Metab 20: 369–380
64. Sullivan HG, Martinez J, Becker DP, Miller JD, Griffith R, Wist AO (1976) Fluid-percussion model of mechanical brain injury in the cat. J Neurosurg 45: 521–534
65. Todd MM, Weeks JB, Warner DS (1993) A focal cryogenic brain lesion does not reduce the minimum alveolar concentration for halothane in rats. Anesthesiology 79: 139–143
66. Vinas FC, Dujovny M, Hodgkinson D (1995) Early hemodynamic changes at the microcirculatory level and effects of mannitol following focal cryogenic injury. Neurol-Res 17: 465–468
67. Waxweiler RJ, Thurman D, Sniezek J, Sosin D, O'Neil J (1995) Monitoring the impact of traumatic brain injury: a review and update. J Neurotrauma 12: 509–516
68. Wilson JT, Gross CE, Bednar MM, Shackford SR (1995) U83836E reduces secondary brain injury in a rabbit model of cryogenic trauma. J Trauma 39: 473–477
69. Yamamura H, Hiraide A, Matsuoka T, Tanaka H, Shimazu T, Sugimoto H (2000) Effect of growth hormone on brain oedema caused by a cryogenic brain injury model in rats. Brain Inj 14: 669–676

Correspondence: R. R. Leker MD, Department of Neurology, Hadassah University Hospital, Ein Kerem, P.O. Box 12000, Jerusalem 91120, Israel.

Acta Neurochir (2002) [Suppl] 83: 55–59

Experimental Models in Focal Cerebral Ischemia: Are we there yet?

R. R. Leker[1] and **S. Constantini**[2]

[1] Department of Neurology and the Agnes Ginges Center for Human Neurogenetics, Hadassah University Hospital Jerusalem, Israel
[2] Department of Pediatric Neurosurgery, Tel Aviv Surasky Medical Center and Tel Aviv University, Tel Aviv, Israel

Summary

Therapeutic options available for acute stroke management are sparse and inadequate. Therefore, new insights into stroke pathophysiology leading to new therapeutic targets are needed. In order to attain these goals, adequate animal models for cerebral ischemia are needed. In the following paper the authors will review the various animal models for stroke and emphasize their potential strengths and weaknesses.

Keywords: Experimental models; stroke; animal; cerebral ischemia.

Introduction

Stroke is the third leading cause of death and the leading cause of chronic disability in Western societies affecting yearly nearly 750000 new cases. Numerous new insights into stroke pathophysiology have been gained over the past years. The involvement of excitatory amino acids [12, 27, 30], free radicals [42], inflammation [1, 4, 16, 17] and apoptosis [13, 14, 29, 34] in damaging ischemic neurons and the involvement of anti-inflammatory cytokines [46] and heat shock proteins [39, 45] in protecting neurons are but a few pertinent examples. Most if not all these new data were obtained from the use of experimental stroke models, which are performed almost exclusively in rodents. Unfortunately, despite all the breakthroughs in the understanding of cerebral ischemia, relatively few therapeutic options are currently available for stroke patients. Therefore, it is imperative that new therapeutic strategies will be devised. For that purpose, adequate animal models of cerebral ischemia are needed in order to learn more about stroke pathophysiology and translate this knowledge into drug development. Stroke animal models are also used to test the safety and efficacy of new pharmaceuticals directed at reducing neuronal mortality. Animals subjected to stroke can be studied for sensory motor [9], behavioral [41] and cognitive [25] abnormalities and their brains can be studied with molecular and histological techniques in order to assess stroke mechanisms and infarct volumes. Much has been said before about the validity of these models to human stroke research [20–22, 38, 49, 54], and it is not the scope of this review to reassess this data. Rather, this review will attempt to present the various animal models for stroke and delineate the strength and weaknesses of each individual model. Experimental stroke models are divided into those of permanent and transient ischemia which both will be discussed in this review.

Intraluminal Filament Occlusion of the Middle Cerebral Artery (MCA)

This model was first developed by Longa *et al.* [35] and is probably the most widely used experimental stroke model. It entails isolation of the common, external and internal carotid arteries through a midline neck dissection. The pterigopalatine branch is ligated and a loose suture is placed over the common carotid. The distal external carotid artery is then ligated. A 4-0 nylon monofilament suture with its tip rounded by heating near a flame is inserted into the external carotid artery through a small arteriotomy and advanced for 19–21 mm into the internal carotid until occluding the proximal middle cerebral artery. Leaving the filament within the artery results in permanent middle cerebral artery occlusion (PMCAO). Application of this model leads to the development of large infarcts involving cortical and subcortical zones (~35–40% hemispheric volume). The model can be used in rats [2,

Table 1. *Characteristics of Experimental Stroke Models*

	Intraluminal filament	Craniotomy electrocoagulation	Auto-embolism	Photo-thrombotic	Endothelin injection
Species	mice/rats	mice/rats	mice/rats	mice/rats	rats
Technical difficulty	moderate	moderate	moderate	mild	high
Reproducibility	mild-moderate	high	moderate	high	unknown
Availability of a reversible ischemia variant	yes	yes (labor intensive)	yes (with thrombolytic drugs)	no	no
Similarity to human pathophysiology	moderate	moderate	high	moderate	low

19] and mice [7, 15] as well as in larger species and it can also be used in MRI machines [33] offering the possibility to evaluate the immediate effects of infarction (Table 1).

This model has several drawbacks. First, the reproducibility of this model is only around 67%–74% [43] with variable results reported among different rat strains [2, 19] and even in the same strain obtained from different breeders [40]. Using this model one assumes to occlude the proximal MCA. However, because the actual occlusion of the MCA is not done under vision, the suture sometimes remains in the distal internal carotid or even enters the proximal anterior cerebral artery (ACA). This often leads to tears in the ACA and to significant subarachnoid or intracerebral hemorrhage [43]. Modifications of this model with coating of the occluding filament with either silicon [2, 47] or poly-L-lysine [7] in order to increase the chances of thrombus formation does appear to increase the reproducibility to around 85%. Use of laser doppler probes to ensure occlusion of the artery may also increase the reproducibility [43] as does brief neurological testing of the animals following awakening from anesthesia and using only the animals with a sensory-motor deficit in the study [8]. Second, the mortality rate of these animals is also very high due to the large volume of the infarct and accompanying edema [43]. This results in the need to increase the number of animals in a given study.

However, this model also has several important advantages. It is rather simple and significantly less invasive than the comparable PMCAO models that use craniotomies. The infarcts typically involve the frontal, parietal and temporal lobes including subcortical zones thereby allowing for more elaborate tests of memory and behavior [2, 7, 19, 24]. Furthermore, since it is possible to retract the occluding filament, one is able to use this model for transient occlusion of the MCA (TMCAO) [15, 41]. Indeed, this model serves as the prototype of TMCAO models. TMCAO models allow for the concomitant use of thrombolytic agents and neuroprotective drugs and also mimic the physiological time course of many strokes in which reperfusion does occur [22]. However, it needs to be remembered that in most human cases the actual infarct is much smaller due to a more distal site of obstruction in the MCA and therefore, it can not be maintained that this model is more representative of the human pathology compared with others.

Permanent Middle Cerebral Occlusion (PMCAO) by Craniotomy and Electrocoagulation

This model was first devised by Bederson *et al.* [6] and was later slightly modified [3]. It entails placement of the animals in a tiltable stereotaxic head holder under a surgical microscope. The temporozygomatic suture is exposed after dissection of the temporalis muscle. A 0.5 mm burr hole is drilled with a dental foot-operated hand held drill and the middle cerebral artery is then exposed all the way down to its junction with the inferior cerebral vein. Then the dura is reflected off the brain. Finally the artery is occluded and cut by electrocoagulation as it is carefully traced off the brain surface.

The infarcts typically involve the fronto-parietal cortex and do not extend into the subcortical zones. They usually involve around 20% of the hemispheric volume [31].

On the up side it could be said that despite the variability in the results obtained with this model in different rat strains its use in spontaneously hypertensive rats (SHR) gives rise to highly reproducible cortical infarcts. Areas of ischemic core and penumbra can be easily discerned with this model, and the effects of different interventions on the fate of these zones can be studied. On the down side, this model is invasive and labor intensive and necessitates excellent surgical skills in order to avoid damage to cortex while exposing the artery and reflecting the dura off of the brain. Furthermore, since it does not usually involve the temporal lobe, the animals can not be adequately tested for memory dysfunctions.

Variants of the original model entail clipping of the MCA (with or without clipping of the internal carotid artery) by an arterial clip [10] or closing the artery with a suture after its exposure. The pros and cons of these modifications are basically similar to that of the original model although it can be argued that they add the element of a foreign body been placed on the brain's surface. Another variant of this model entails pulling the MCA off the brain and hanging it on a rod so that a kink is formed in the artery preventing blood flow. At a later time point one can lower the artery back to its natural place so the kink in the artery is removed and blood flow to the ischemic brain is restored [5]. This TMCAO variant is labor intensive and great care needs to be taken to ensure that the flow is indeed blocked distal to the kink and that the artery is not damaged and torn by the pull. Therefore, this technique for TMCAO is seldom used.

The Auto-Embolic Model

This model is the most recent experimental stroke model devised and can be used in rats [51] and mice [52] or rabbits [32]. It entails preparation of an autologous venous blood clot. This primary clot is then cut into microembolic particles, which are injected intra-arterially into the distal internal carotid artery or proximal MCA resulting in single or multiple embolic infarcts depending on the number and size of injected micro-emboli [51].

This model has several weak points: The reproducibility of the infarcts attained with this model is not so high and depends on the size of the microemboli. Furthermore, since the clots are injected into the MCA or ICA, they occlude distal portions of the MCA or ACA and thus the size of the infarct and its location are subject to variations [36]. As in the suture model one can improve the reproducibility by using laser doppler flow monitors as well as motor disability scores to eliminate from the study those animals with no signs of reduced perfusion or disability. It is also labor intensive and time consuming.

On the brighter side, this model is somewhat more physiological than others and simulates arterial occlusion with true endogenous clot. It also allows for the administration of thrombolytic agents together with other protective measures [11].

Several modifications of this model have been published lately including the injection of artificial microembolic particles into the MCA [11, 28]. This allows for better control of the size of the embolus but leads to scattered infarctions due to the fact that multiple particles are usually injected [18].

A further modification of this model entails intraarterial injection of thrombin to the distal internal carotid or proximal MCA [53]. This leads to the formation of a clot at the site of injection and to infarction. This model has not been studied extensively thus far and therefore, its comparison to other models can not be assessed at this time.

Photothrombotic Occlusion of the MCA

This model involves systemic administration of the Rose Bengal dye (or other photosensitizing agents) [37] and irradiating the exposed segment of the MCA with either green light [37] at a certain wave length or with a laser beam [23, 50]. This gives rise to endothelial damage, thrombosis and occlusion of the MCA. The model appears to be highly reproducible but is not frequently used today. It usually causes a well-delineated cortical infarct [23] similarly to the electrocoagulation model described above and has the same advantages and disadvantages as that model.

Intracerebral Injection of Endothelins

This model involves direct intracerebral injection of the highly potent vasoconstrictors endothelin1 or endothelin3 [26, 44]. Both agents cause severe vasospasm and eventually thrombosis of the adjacent arteries. This results in irreversible ischemia. This model is seldom used because it is highly invasive and because direct brain injury can be caused by the intracerebral injections.

Small Vessel Occlusion by Sodium Laurate

This model was recently published by Toshima *et al.* [48]. It entails injection of small quantities of sodium laurate intra-arterially. The injection causes endothelial damage in small cerebral vessels and leads to thrombus formation composed of fibrin, platelets, red blood cells and white blood cells in these vessels [48]. This results in multiple small infarcts involving the hippocampus, cortex and thalamus [48]. Due to the novelty of this model, its reproducibility is not yet clear and therefore it can not be directly compared with other models. Nevertheless, it clearly is a welcomed addition to the arsenal of stroke investigators in that it adds a model for small vessels disease, an area that is neglected by researchers despite its high frequency in human stroke patients.

In conclusion, there are many models of focal cerebral ischemia that are currently available to researchers in the field. Each model has its weaker and stronger points and no single model is categorically superior to others in mimicking the actual behavior of a typical human infarct. Furthermore, stroke is not a homogenous disorder but rather each stroke subtype (e.g. small vessel disease versus cardioembolic stroke) involves different pathogenetic mechanisms and pathways. Therefore, in order to better understand stroke pathophysiology and discover new and effective therapeutic strategies to combat stroke, one would probably need to treat each hypothesis in different stroke models. Nevertheless, stroke models are indispensable for improving the future of stroke care.

References

1. Arvin B, Neville LF, Barone FC, Feuerstein GZ (1996) The role of inflammation and cytokines in brain injury. Neurosci Biobehav Rev 20: 445–452
2. Aspey BS, Cohen S, Patel Y, Terruli M, Harrison MJ (1998) Middle cerebral artery occlusion in the rat: consistent protocol for a model of stroke. Neuropathol Appl Neurobiol 24: 487–497
3. Barone FC, Arvin B, White RF, Miller A, Webb CL, Willette RN, Lysko PG, Feuerstein GZ (1997) Tumor necrosis factor-alpha. A mediator of focal ischemic brain injury. Stroke 28: 1233–1244
4. Barone FC, Feuerstein GZ (1999) Inflammatory mediators and stroke: new opportunities for novel therapeutics. J Cereb Blood Flow Metab 19: 819–834
5. Barone FC, Feuerstein GZ, White RF (1997) Brain cooling during transient focal ischemia provides complete neuroprotection. Neurosci Biobehav Rev 21: 31–44
6. Bederson JB, Pitts LH, Tsuji M, Nishimura MC, Davis RL, Bartkowski H (1986) Rat middle cerebral artery occlusion: evaluation of the model and development of a neurologic examination. Stroke 17: 472–476
7. Belayev L, Busto R, Zhao W, Fernandez G, Ginsberg MD (1999) Middle cerebral artery occlusion in the mouse by intraluminal suture coated with poly-L-lysine: neurological and histological validation. Brain Res 833: 181–190
8. Borlongan CV, Hida H, Nishino H (1998) Early assessment of motor dysfunctions aids in successful occlusion of the middle cerebral artery. Neuroreport 9: 3615–3621
9. Borlongan CV, Randall TS, Cahill DW, Sanberg PR (1995) Asymmetrical motor behavior in rats with unilateral striatal excitotoxic lesions as revealed by the elevated body swing test. Brain Res 676: 231–234
10. Buchan AM, Xue D, Slivka A (1992) A new model of temporary focal neocortical ischemia in the rat. Stroke 23: 273–279
11. Busch E, Kruger K, Hossmann KA (1997) Improved model of thromboembolic stroke and rt-PA induced reperfusion in the rat. Brain Res 778: 16–24
12. Choi D (1998) Antagonizing excitotoxicity: a therapeutic strategy for stroke? Mt Sinai J Med 65: 133–138
13. Choi DW (1996) Ischemia-induced neuronal apoptosis. Curr Opin Neurobiol 6: 667–672
14. Chopp M, Li Y (1996) Apoptosis in focal cerebral ischemia. Acta Neurochir (Wien) [Suppl] 66: 21–26
15. Clark WM, Lessov NS, Dixon MP, Eckenstein F (1997) Monofilament intraluminal middle cerebral artery occlusion in the mouse. Neurol Res 19: 641–648
16. DeGraba TJ (1998) The role of inflammation after acute stroke: utility of pursuing anti-adhesion molecule therapy. Neurology 51: S62–68
17. del Zoppo G, Ginis I, Hallenbeck JM, Iadecola C, Wang X, Feuerstein GZ (2000) Inflammation and stroke: putative role for cytokines, adhesion molecules and iNOS in brain response to ischemia. Brain Pathol 10: 95–112
18. Demura N, Mizukawa K, Ogawa N, Yamashita K, Kanazawa I (1993) A cerebral ischemia model produced by injection of microspheres via the external carotid artery in freely moving rats. Neurosci Res 17: 23–30
19. Dogan A, Baskaya MK, Rao VL, Rao AM, Dempsey RJ (1998) Intraluminal suture occlusion of the middle cerebral artery in spontaneously hypertensive rats. Neurol Res 20: 265–270
20. Ginsberg MD (1996) The validity of rodent brain-ischemia models is self-evident [see comments]. Arch Neurol 53: 1065–1067
21. Ginsberg MD, Busto R (1989) Rodent models of cerebral ischemia. Stroke 20: 1627–1627
22. Grotta J (1996) Rodent models of stroke limitations. What can we learn from recent clinical trials of thrombolysis? [see comments]. Arch Neurol 53: 1067–1070
23. Gu WG, Brannstrom T, Jiang W, Wester P (1999) A photothrombotic ring stroke model in rats with remarkable morphological tissue recovery in the region at risk. Exp Brain Res 125: 171–183
24. Hata R, Mies G, Wiessner C, Fritze K, Hesselbarth D, Brinker G, Hossmann KA (1998) A reproducible model of middle cerebral artery occlusion in mice: hemodynamic, biochemical, and magnetic resonance imaging. J Cereb Blood Flow Metab 18: 367–375
25. Hattori K, Lee H, Hurn PD, Crain BJ, Traystman RJ, DeVries AC (2000) Cognitive deficits after focal cerebral ischemia in mice. Stroke 31: 1939–1944
26. Henshall DC, Butcher SP, Sharkey J (1999) A rat model of endothelin-3-induced middle cerebral artery occlusion with controlled reperfusion. Brain Res 843: 105–111
27. Hossmann KA (1994) Glutamate-mediated injury in focal cerebral ischemia: the excitotoxin hypothesis revised. Brain Pathol 4: 23–36
28. Kilic E, Hermann DM, Hossmann KA (1998) A reproducible model of thromboembolic stroke in mice. Neuroreport 9: 2967–2970
29. Kogure T, Kogure K (1997) Molecular and biochemical events within the brain subjected to cerebral ischemia (targets for therapeutical intervention). Clin Neurosci 4: 179–183
30. Lee JM, Zipfel GJ, Choi DW (1999) The changing landscape of ischaemic brain injury mechanisms. Nature 399: A7–14
31. Leker RR, Shohami E, Abramsky O, Ovadia H (1999) Dexanabinol; a novel neuroprotective drug in experimental focal cerebral ischemia. J Neurol Sci 162: 114–119

32. Lew SM, Gross CE, Bednar MM, Russell SJ, Fuller SP, Ellenberger CL, Howard D (1999) Complement depletion does not reduce brain injury in a rabbit model of thromboembolic stroke. Brain Res Bull 48: 325–231
33. Li F, Han S, Tatlisumak T, Carano RA, Irie K, Sotak CH, Fisher M (1998) A new method to improve in-bore middle cerebral artery occlusion in rats: demonstration with diffusion- and perfusion-weighted imaging. Stroke 29: 1715–1719; discussion 1719–1720
34. Li W, Jin K, Nagayama T, He X, Chang J, Minami M, Graham SH, Simon RP, Greenberg DA (2000) Increased expression of apoptosis-linked gene 2 (ALG2) in the rat brain after temporary focal cerebral ischemia. Neuroscience 96: 161–168
35. Longa EZ, Weinstein PR, Carlson S, Cummins R (1989) Reversible middle cerebral artery occlusion without craniectomy in rats. Stroke 20: 84–91
36. Lyden PD, Zivin JA, Chabolla DR, Jacobs MA, Gage FH (1992) Quantitative effects of cerebral infarction on spatial learning in rats. Exp Neurol 116: 122–132
37. Matsuno H, Uematsu T, Umemura K, Takiguchi Y, Asai Y, Muranaka Y, Nakashima M (1993) A simple and reproducible cerebral thrombosis model in rats induced by a photochemical reaction and the effect of a plasminogen-plasminogen activator chimera in this model. J Pharmacol Toxicol Methods 29: 165–173
38. Millikan C (1992) Animal stroke models [editorial]. Stroke 23: 795–797
39. Nowak TS Jr, Jacewicz M (1994) The heat shock/stress response in focal cerebral ischemia. Brain Pathol 4: 67–76
40. Oliff HS, Weber E, Miyazaki B, Marek P (1995) Infarct volume varies with rat strain and vendor in focal cerebral ischemia induced by transcranial middle cerebral artery occlusion. Brain Res 699: 329–331
41. Sakai N, Yanai K, Ryu JH, Nagasawa H, Hasegawa T, Sasaki T, Kogure K, Watanabe T (1996) Behavioral studies on rats with transient cerebral ischemia induced by occlusion of the middle cerebral artery. Behav Brain Res 77: 181–188
42. Samdani AF, Dawson TM, Dawson VL (1997) Nitric oxide synthase in models of focal ischemia. Stroke 28: 1283–1288
43. Schmid-Elsaesser R, Zausinger S, Hungerhuber E, Baethmann A, Reulen HJ (1998) A critical reevaluation of the intraluminal thread model of focal cerebral ischemia: evidence of inadvertent premature reperfusion and subarachnoid hemorrhage in rats by laser-Doppler flowmetry. Stroke 29: 2162–2170
44. Sharkey J, Butcher SP (1995) Characterisation of an experimental model of stroke produced by intracerebral microinjection of endothelin-1 adjacent to the rat middle cerebral artery. J Neurosci Meth 60: 125–131
45. Soriano MA, Planas AM, Rodriguez Farre E, Ferrer I (1994) Early 72-kDa heat shock protein induction in microglial cells following focal ischemia in the rat brain. Neurosci Lett 182: 205–207
46. Spera PA, Ellison JA, Feuerstein GZ, Barone FC (1998) IL-10 reduces rat brain injury following focal stroke. Neurosci Lett 251: 189–189
47. Takano K, Tatlisumak T, Bergmann AG, Gibson DG 3rd, Fisher M (1997) Reproducibility and reliability of middle cerebral artery occlusion using a silicone-coated suture (Koizumi) in rats. J Neurol Sci 153: 8–11
48. Toshima Y, Satoh S, Ikegaki I, Asano T (2000) A new model of cerebral microthrombosis in rats and the neuroprotective effect of a Rho-kinase inhibitor. Stroke 31: 2245–2250
49. Wiebers DO, Adams HP Jr, Whisnant JP (1990) Animal models of stroke: are they relevant to human disease? [editorial] [see comments]. Stroke 21: 1–3
50. Yao H, Ibayashi S, Sugimori H, Fujii K, Fujishima M (1996) Simplified model of krypton laser-induced thrombotic distal middle cerebral artery occlusion in spontaneously hypertensive rats. Stroke 27: 333–336
51. Zhang RL, Chopp M, Zhang ZG, Jiang Q, Ewing JR (1997) A rat model of focal embolic cerebral ischemia. Brain Res 766: 83–92
52. Zhang Z, Chopp M, Zhang RL, Goussev A (1997) A mouse model of embolic focal cerebral ischemia. J Cereb Blood Flow Metab 17: 1081–1081
53. Zhang Z, Zhang RL, Jiang Q, Raman SB, Cantwell L, Chopp M (1997) A new rat model of thrombotic focal cerebral ischemia. J Cereb Blood Flow Metab 17: 123–135
54. Zivin JA, Grotta JC (1990) Animal stroke models. They are relevant to human disease. Stroke 21: 981–983

Correspondence: R. R. Leker M.D., Department of Neurology, Hadassah University Hospital, Ein Kerem, P.O. Box 12000, Jerusalem 91120, Israel.

Acta Neurochir (2002) [Suppl] 83: 61–69

Experimental Subarachnoid Haemorrhage Models in Rats

T. Alkan[1], **E. Korfali**[2], and **N. Kahveci**[1]

[1] Department of Physiology, Uludag University, School of Medicine , Bursa, Turkey
[2] Department of Neurosurgery, Uludag University, School of Medicine, Bursa, Turkey

Summary

There is no comprehensive and reliable model available in small animals that are suitable for the study of subarachnoid haemorrhage (SAH). In the study we reviewed the advantages and disadvantages of available SAH models in rats and presented our model.

Experimental SAH was induced in a group of 350–450 g Sprague-Dawley rats. A 2 mm-diameter burr hole was drilled and, working under a microscope, haemorrhage was produced by transclival puncture of the basilar artery with a 20 μm thick piece of glass. The rats were assigned to either the experimental group (n: 7) or the control group (n: 7). Local cerebral blood flow (LCBF), intracranial pressure (ICP), and cerebral perfusion pressure (CPP) were measured for 60 min after SAH, after which the rats were decapitated. Microscopic examinations were done on three different segments of the basilar artery.

There was a significant and sharp drop in LCBF just after SAH was induced (56.17 ± 12.80 mlLD/min/100 g and 13.57 ± 5.85 mlLD/min/100 g for baseline and post-SAH, respectively; $p < 0.001$), the flow slowly increased by the end of the experiment but never recovered to pre-SAH values ($43{,}63 \pm 7.6$ mlLD/min/100 g, $p < 0.05$). ICP (baseline 7.33 ± 0.8 mmHg) increased acutely to 70.6 ± 9.2 mmHg, and also returned to normal levels by 60 min after SAH. CPP (baseline 75.1 ± 4.9 mmHg) dropped accordingly (to 21.0 ± 6.3 mmHg) and then increased, reaching 70.1 ± 4.9 mmHg at 60 min after SAH. Examinations of the arteries revealed decreased inner luminal diameter and distortion of the elastica layer.

We present an inexpensive and reliable model of SAH in the rat that allows single and multiple haemorrhages and to study the early and late course of pathological changes.

Keywords: Acute vasoconstriction; cerebral blood flow; animal models; intracranial pressure; subarachnoid haemorrhage.

Introduction

The decreased local cerebral blood flow (LCBF) and cerebral ischemia that occur after subarachnoid haemorrhage (SAH) may be caused by acute and/or delayed vasospasm.

One of the most important and critical aspects of SAH-induced vasospasm is its failure to consistently respond to treatment. Pharmacological interventions have been tried in experimental models and in clinical trials with only partial success. Because the experimental study of cerebral arteries in live humans are not possible, animal models have been developed [3, 5, 7, 10, 14, 16, 24, 29, 30, 43, 55, 62, 64]. In this article, we discussed the advantages and disadvantages of available SAH models in rats and presented our model.

Materials and Methods

All the procedures and protocols were reviewed and approved by the Animal Care and Use Committee at Uludağ University School of Medicine. Fourteen adult male Sprague-Dawley rats weighing 350–450 g were assigned to either the experimental group (n: 7) or the control group (n: 7). The animals were fasted overnight but allowed free access to water. Each rat was anaesthetised in anaesthesia chamber (EJAY Anaesthesia Chamber, International Inc., California, USA) with 1% isoflurane, transorally intubated, and mechanically ventilated (Small Animal Ventilator, Model 683, Harvard Apparatus Inc., MA, USA) with a mixture of 30% O_2 + 70% N_2O. Anaesthesia was maintained with 0.5–1% isoflurane, with settings adjusted to maintain arterial $PaC0_2$ at 30–40 mmHg and partial pressure of oxygen at 100–150 mmHg. An intraperitoneal injection of 0.08-mg/kg atropine sulphate was given to reduce tracheal secretions. The left femoral artery was catheterised with PE-50 polyethylene tubing for blood sampling and continuous recording of arterial blood pressure. The left femoral vein was catheterised with the same type of device, and drugs and fluids were administered through this access. Arterial blood samples were drawn using micro capillary tubes, and pH, $PaCO_2$, and PaO_2 were measured prior to the induction of SAH and at the end of the experiment using a blood gas analyser (Chiron Diagnostics, Model 348, UK). Separate samples were drawn for analysis of blood glucose levels using Glucostix reagent strips and a reflectometer system (Medisense Inc., Waltham, MA, USA). Body temperature was monitored with a rectal probe (Protocol systems Inc., OR, USA), and was maintained at 37.0 ± 0.5°C using a heating pad.

Once anaesthetised and catheterised, each rat was placed in the prone position. In order to measure ICP, a 1-cm midline incision was made starting just behind the occipital bone and extending along the length of the first and second cervical vertebrae. When the atlanto-occipital membrane was exposed, a cannula attached to a pressure

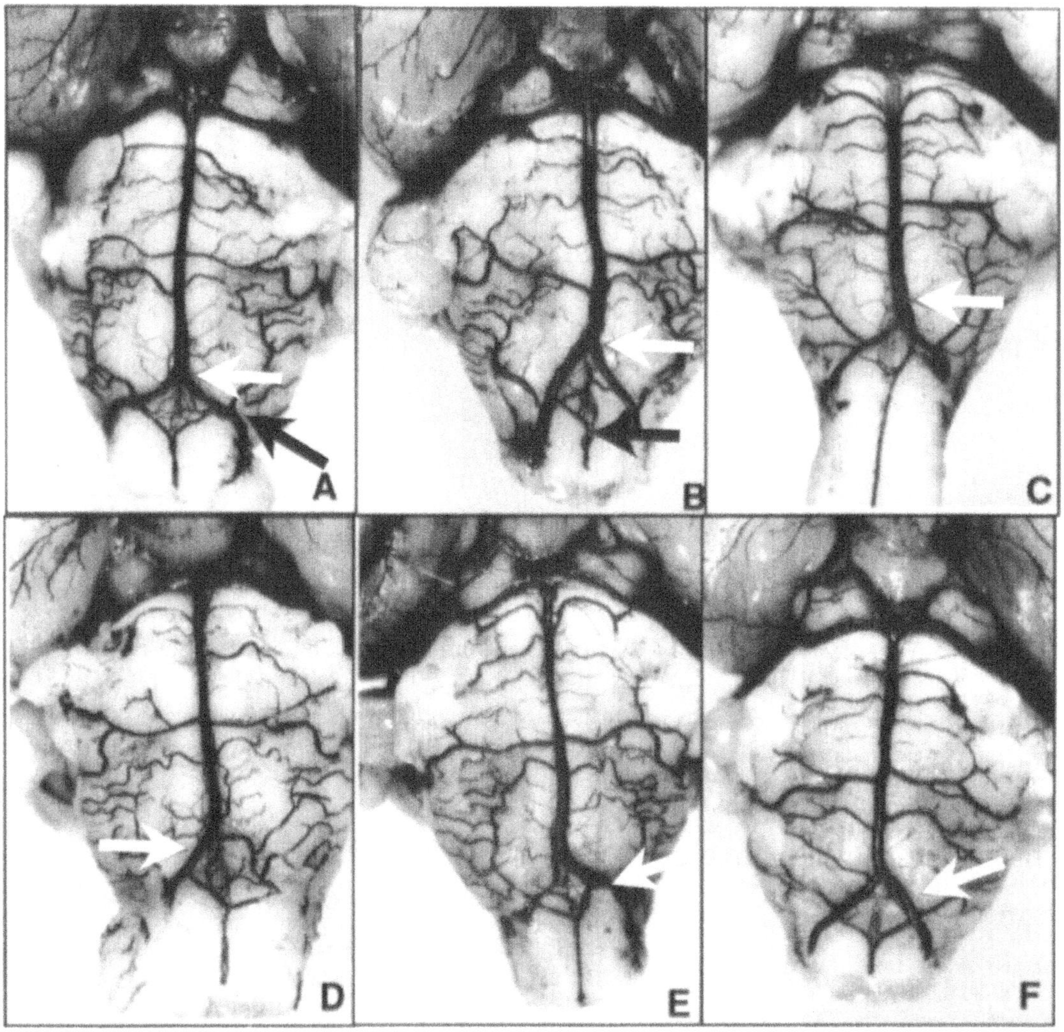

Fig. 1. Anatomical variations of vertebro-basilar arteries in rats. (A) Normal variation (68%) (White arrow: Normal junction site of vertebral arteries; Black arrow: Posterior inferior cerebellar artery). (B) High bifurcation of vertebral arteries (32%) (White arrow), Anterior spinal artery (Black arrow). (C) Forking of the basilar artery (1%) (White arrow). (D) Dominant right vertebral artery (24%) (White arrow). (E) Dominant left vertebral artery (6%). (F) Two seperate vertebral arteries (5%) (White arrow) (Adapted from Alkan and Kahveci, 1999)

transducer was inserted into the subarachnoid space and positioned such that a satisfactory ICP waveform was obtained, according to the method described by Barth and colleagues [6]. Once the cannula was in place, methlymethacrylate cement was used to fix it to a stainless steel screw that had been implanted in the occipital bone. When the ICP waveform on the monitor was satisfactory, the rat was shifted to the supine position. After all the recording devices were properly connected, the animal was positioned in a stereotactic frame and kept there for the remainder of the experiment.

We modified Barry *et al.* [5] transclival approach for producing SAH in our experiment. Briefly after topical anaesthetic (10% Xylocaine Spray, Astra Södertalje, Sweden) was applied to the surgery site, a 2-cm midline incision was made in the anterior neck and the following procedures were performed using a surgical microscope (Carl Zeiss, 99, Germany). The sternohyoid and omohyoid muscles were blunt dissected and separated, and the trachea and oesophagus were gently retracted using a hand-held retractor. This exposed the base of the skull and the insertion of the longus capitis muscle. A 2 mm-diameter burr hole was made at this site using a saline-cooled electrical high-speed drill (Aesculap Microton GD412, Germany) equipped with a diamond burr, with the depth such that only a thin translucent cranial plate remained. Under this thin layer of bone we were able to observe the positions of the vertebral arteries and their branches, including the basilar artery, and the posterior inferior and anterior inferior cerebellar arteries. All animals with anatomical variations of the basilar artery were excluded and only the group A rats were used [1] (Fig. 1). A separate 1 mm-diameter burr hole was made just above the first hole, without penetrating the dura. Into this we placed a laser Doppler flowmetry probe (0.8 mm diameter, Model P-433, Vasomedics Inc., MN, USA), which was advanced to the epidural surface under stereotactic control and used to measure LCBF.

In the experimental group, once the positions of the arterial branches were observed through the initial 2 mm-diameter opening, a 23 G needle was used to make a small hole in the paper-thin inner table of the bone. We then punctured the basilar artery with a 20 µm-thick glass tip and immediately sealed the bone defect with wax, without allowing any cerebrospinal fluid leakage. Each animal was

Table 1. *The Physiological Findings in the Groups*

	pH	$PaCO_2$ (mmHg)	PaO_2 (mmHg)	Hematocrit (%)	Glucose (mmol/L)	Rectal temperature (°C)
Control group (n: 7)	7.41 ± 0.05	42.3 ± 3.1	103.1 ± 8.0	41.0 ± 2.3	5.7 ± 0.5	37.4 ± 0.4
Experimental group (n: 7)	7.37 ± 0.04	38.6 ± 4.2	109.8 ± 10.5	39.4 ± 1.3	4.7 ± 0.7	37.1 ± 0.2

continuously monitored for 30 min before and 60 min after SAH, and cerebral hemodynamic values (ICP, mean arterial blood pressure [MABP], CPP = MABP – ICP, and LCBF) were recorded every 5 minutes. LCBF expressed as a percentage of baseline. The rats in the control group underwent the same procedures as those in the experimental group except for SAH induction.

After all monitoring was completed, in order to assess the acute effects of SAH on the basilar artery, all the rats were transcardially perfused with saline and 4% buffered paraformaldehyde, and their brains were removed and fixed in 4% paraformaldehyde. Using an operating microscope, three segments of the artery were dissected from each brain: 0.5 mm caudal to the vertebrobasilar junction (section v); 0.5 mm rostral to the junction with the anterior inferior cerebellar artery (section a); and 0.5 mm rostral to the junction with the posterior cerebral artery (section p). Ten micron-thick sections were cut from each of these lengths and were prepared with haematoxylin-eosin and Masson's trichrome stains. The slides were examined for histopathological changes, and vessel wall thickness, inner luminal diameter, and media thickness were measured. The vessel measurements were made by the same pathologist, who was unaware of the group to which each specimen belonged.

The data was statistically analysed using ANOVA and the Tukey-Kramer multiple comparisons test. $p < 0.05$ was considered significant.

Results

There were no significant differences between the control and experimental groups regarding arterial pH, $PaCO_2$, PaO_2, hematocrit, blood glucose levels, or body temperature before and after SAH (Table 1).

SAH was associated with an immediate brief rise in MABP (100.5 ± 8.4 mmHg), but the values quickly returned to baseline values (82.4 ± 9.9 mmHg) ($p > 0.01$). The ICP at baseline was 7.33 ± 0.8 mmHg, but this spiked to 70.6 ± 9.2 mmHg immediately after SAH ($p < 0.001$) and then gradually fell, returning to just above baseline by 60 min after SAH (10 ± 0.8 mmHg) ($p > 0.05$). There was a significant and sharp drop in LCBF just after SAH induction (56.17 ± 12.80 mlLD/min/100 g baseline versus 13.57 ± 5.85 mlLD/min/100 g post-SAH; $p < 0.001$), 20 minutes later the flow began to increase but never recovered to pre-SAH values (43.63 ± 7.6 mlLD/min/100 g; $p < 0.05$). CPP also declined immediately after SAH, dropping from 75.1 ± 4.9 mmHg to 21.0 ± 6.3 mmHg ($p < 0.001$) and then gradually recovering to near baseline by the 60-min stage (70.1 ± 4.9 mmHg) ($p > 0.05$) (Fig. 2).

When the tissues were dissected for histological preparation, we noted that the experimental rats had a thick subarachnoid clot over the basal surface of the brain in the region of the basilar artery and the major cerebral arteries. Microscopic study revealed that the basilar arteries in the seven brains from the experimental group were moderately to severely constrict. The mean inner luminal diameter of the experimental animals' vessels was 57.6% smaller after SAH than the diameter in the control group (v section: $p < 0.01$, a and p sections: $p < 0.001$) (Table 2). Also, vessel wall thickness was significantly greater in the group that sustained SAH compared to the controls (Table 3). Regarding measurements of basilar artery medial wall thickness, the puncture site was the only segment that was significantly thinner in the experimental rats (Table 4) ($p < 0.05$). The only remarkable histological change noted in the experimental group was distortion of the elastica layer (5/7 rats). Only one rat in the control group showed minimal elastica distortion.

Discussion

It is interesting that so many different animal models of SAH have been developed in so many different species. In vivo experimental models have been used to investigate diverse aspects of vasospasm, including its natural history [4, 44, 45], pathogenesis [36, 37, 40, 41], pathology [12, 20, 22], diagnosis [5], and treatment. A model used to investigate one aspect may not be suitable for another. In addition, some models require special equipment and trained personnel. Cost and availability of animals may also be an important factor.

The development of reliable and reproducible animal models of SAH is required for the systematic study of the pathophysiology and treatment of vasospasm. Desirable models are those that replicate features of human SAH. Various species have been used including dogs [3, 30], cats [29, 62], monkeys [10, 16, 26, 29, 46,

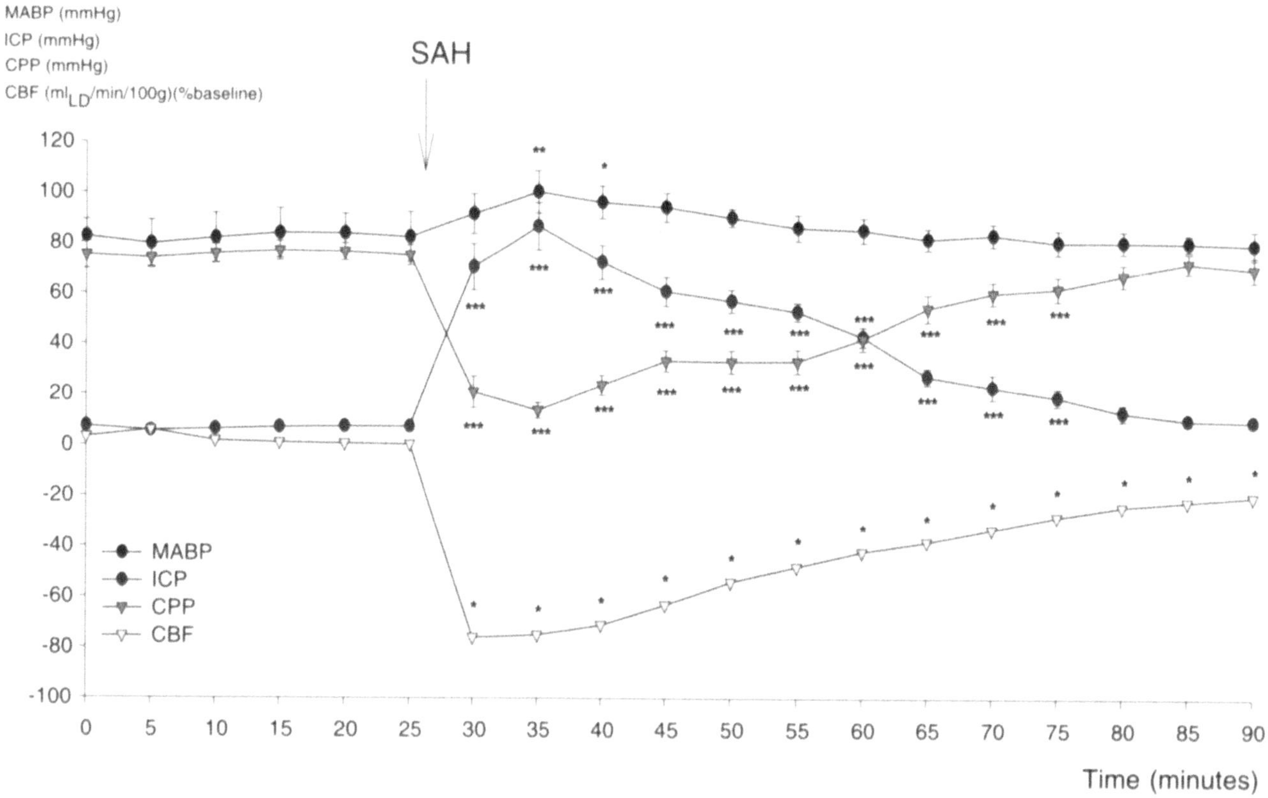

Fig. 2. The pre and post subarachnoid haemorrhage measurements of MABP, ICP, LCBF and CPP in the experimental group (CBF was expressed as % of baseline). *MABP* Mean arterial blood pressure, *ICP* Intracranial blood pressure, *LCBF* Local cerebral blood flow, *CPP* Cerebral perfusion pressure *$p < 0.05$, **$p < 0.01$, ***$p < 0.001$ compared to baseline values

Table 2. *Basilar Artery Luminal Diameter (Mean ± SD)*

	Section v (μm)	Section a (μm)	Section p (μm)
CONTROL GROUP (n: 7)	112.49 ± 34.50	116.05 ± 21.88	120.98 ± 22.64
Experimental group (n: 7)	64.78 ± 8.85*	62.00 ± 12.67◆	58.42 ± 10.15◆

*: $p < 0.01$, ◆: $p < 0.001$ compared to the control group.
Section v: 0.5 mm caudal to the vertebrobasilar junction.
Section a: 0.5 mm rostral to the junction with the anterior inferior cerebellar artery.
Section p: 0.5 mm rostral to the junction with the posterior cerebral artery.

Table 3. *Basilar Artery Wall Thickness (Mean ± SD)*

	Section v (μm)	Section a (μm)	Section p (μm)
Control group (n: 7)	35.00 ± 3.25	36.00 ± 3.48	33.50 ± 7.70
Experimental group (n: 7)	48.50 ± 8.19*	47.78 ± 10.42•	45.01 ± 4.25*

*: $p < 0.01$, •: $p < 0.05$ compared to the control group.
Section v: 0.5 mm caudal to the vertebrobasilar junction.
Section a: 0.5 mm rostral to the junction with the anterior inferior cerebellar artery.
Section p: 0.5 mm rostral to the junction with the posterior cerebral artery.

Table 4. *Basilar Artery Medial Wall Thickness (Mean ± SD)*

	Section v (μm)	Section a (μm)	Section p (μm)
Control group (n: 7)	28.07 ± 5.25	29.50 ± 4.53	35.50 ± 5.50
Experimental group (n: 7)	36.92 ± 6.23*	36.42 ± 6.54	28.64 ± 7.38

*: $p < 0.05$ compared to the control group.
Section v: 0.5 mm caudal to the vertebrobasilar junction.
Section a: 0.5 mm rostral to the junction with the anterior inferior cerebellar artery.
Section p: 0.5 mm rostral to the junction with the posterior cerebral artery.

50], and more recently rats [5, 7, 14, 15, 64]. The rat is a well-studied, readily available animal that has been widely used and intensively investigated and is preferred for cerebral blood flow studies. A variety of factors, including the drive for improved cost effectiveness has led to the increasing use of the rat as model in the research laboratory. In addition, the rat features important advantages compared with other species as a model for the study of SAH.

Because there are no naturally occurring animal models of vasospasm, a number of techniques have been used to deliberately produce SAH in animals. These techniques generally fall into one of three categories: 1) an artery is punctured allowing blood to escape and collect around the artery and its neighbours; 2) an artery is surgically exposed, and autologous blood is placed around the artery; 3) blood is injected into the subarachnoid space and is allowed to collect around the artery. Each of these techniques has its advantages and disadvantages. The most commonly used method of producing SAH in rats is based on the introduction of autologous blood into the subarachnoid space through the cisterna magna [14, 24, 28, 55]. This technique has several major disadvantages. The model neglects the importance of the injury to the artery in the development of pathologic cerebral hemodynamics. There is evidence that the degree of vasospasm that develops is related to the volume and location of the blood in the subarachnoid space [23, 34]. Furthermore, it does not take into account the importance of the arterial rupture and the immediate pressure effects of the ejected blood to vasospasm. For those groups who use puncture or tearing techniques, it is important to stimulate this aspect of the process. For those who place or inject blood around arteries, the actual presence of a large volume of blood in contact with the vessel is the overriding factor in development of the vasospasm [43]. As demonstrated in numerous studies, the advantage of injecting blood is that it is generally easier to perform and is associated with less animal morbidity and mortality, which may explain its greater popularity.

Barry *et al.* reported on the first use of rats in an experimental model of vasospasm [5]. Anesthetized animals were intubated and, using a surgical microscope, a tungsten microelectrode was positioned stereotactically above the basilar artery through transclival approach and punctured. The degree of vasospasm was subsequently determined by direct visualisation. Vasospasm was significant on the second postoperative day but had resolved on the third day, by which time the blood in the subarachnoid space had also cleared.

In 1995, Bederson *et al.* [7] and Veelken *et al.* [64] published in the same journal almost identical method of introducing intraluminal thread and rupturing internal carotid artery for producing SAH, based on the modified method described for cerebral ischemia by Zea-Longa *et al.* [68]. Although both groups claimed that a non-craniotomy model has several advantages over models that require craniotomy, vessel avulsion, or injection of blood, a potential disadvantage of this model is the brief period of ischemia caused by the intraluminal suture and sacrification of external carotid artery, which also supplies considerable amount of blood to brain through collaterals. Bederson *et al.* [7] claim that the duration of focal ischemia (>4 min) is shorter than required to produce infarction, referring to the work of Kaplan *et al.* [33]. In this model the amount of haemorrhage also cannot be controlled and if required recurrent haemorrhages are impossible. Another and important disadvantage of this model is the incidence of intracerebral haemorrhage (11%) [7], high incidence of mortality rate (>50%) in less than 24 hours after SAH and 11% failure of producing SAH [64]. On the other hand, our model allows controllable amount of haemorrhage with changing the thickness of the tip of glass tube, if desired recurrent haemorrhages precisely from the same bleeding point, topical application of vasodilator agents, removal of clot or substrate collecting around the artery and also closing the burr hole with a glass provides opportunity to observe the arterial changes in a long periodical time (cranial window).

Acute elevation of ICP is one obvious cause of acute ischemia after SAH. Previous experimental studies using intraluminal thread for producing SAH have also demonstrated an immediate rise in ICP and an immediate reduction in CPP, with both parameters returning to baseline within 60 min in most cases [7, 8, 64]. On the other hand Kader *et al.* [31] using the original transclival method described by Barry *et al.* [5] due to the fact that large opening of the clivus and dura were unable to notice increase in ICP. This problem was solved in our method by opening a 23-gauge diameter hole and immediately sealing the hole by bone wax after arterial puncture without allowing any cerebrospinal fluid leakage. Ischemic changes have been attributed solely to this shift in ICP [32]. However, other investigations have revealed a decrease in LCBF after SAH, even under conditions of normal ICP [63]. Bederson and colleagues [8] demonstrated a dissociation of ICP and LCBF during the acute phase of SAH. They questioned whether other factors, such as acute vasoconstriction, might contribute to cerebral ischemia, independent of ICP elevation.

Numerous investigators have demonstrated acute cerebral vasoconstriction as a consequence of experimental SAH [14, 16]. In addition, local cerebral blood flow (LCBF) is known to decrease immediately after SAH, resulting in ischemic injury [8, 29, 35, 48]. Studies have shown that cerebral arteries respond to SAH in a biphasic pattern of constriction. Acute vasoconstriction starts minutes after the haemorrhage begins, and a subsequent phase of delayed vasospasm commences 48 hours later [14]. The significance of delayed vasospasm is recognized [54], but the ways in which acute vasoconstriction contributes to ischemic brain damage after SAH are not well understood [8, 51].

After SAH, there is usually a biphasic change in the calibre of the cerebral blood vessels [11, 14]. In the acute phase of this type of haemorrhage, the blood vessels narrow due to constriction of the smooth-muscle layers. In our study, we observed acutely developed vasoconstriction with the decreased inner lamina diameter, increased vessel wall thickness, and intimal distortion in the early stage after SAH. Other authors have reported similar findings, including vessel wall thickening and increased corrugation of the intima [8, 52]. However, several days after SAH, secondary constriction of the blood vessels has been documented [14]. In our laboratory we also observed this delayed vasoconstriction [2], which appears to be due to an irreversible change in vessel wall structure, not merely altered smooth-muscle tone as suggested by Bevan *et al.* [9].

In our above mentioned study we observed that LCBF was slightly lower than normal values at 4 days post-haemorrhage, with marked histological changes of endothelial distortion, adventitial inflammation, and distortion of the elastica layer. Although the inner luminal diameter of the basilar artery was increased compared to its size in the acute phase, the artery lumen in the experimental rats at 4 days after SAH was still narrower than that in control animals. Our findings are in accord with those of other authors [9, 28, 47, 53, 58, 67].

In contrast, some other investigators of SAH have reported no such vasculopathic changes [31, 57]. They suggested that significant quantities of escaped blood had not remained in contact with the vessels long enough to produce vasospasm, and also attributed their findings to the lack of intra-adventitial spaces in rats' arterial walls. It has been demonstrated that haemoglobin has preferential vasoconstrictive activity in cerebral arterial smooth muscle [18, 61], and a variety of substrates and processes have been implicated in vasoconstriction after SAH. These include haemoglobin or whole blood [19], angiotensin [48], free-radical production [59], catecholamine release [15], vasopressin [13], endothelins [66], and nitric oxide (NO) [21, 27].

Regarding the latter, SAH is associated with decreased local availability of NO, which means that NO-induced cerebrovascular relaxation is impaired [60]. Recently, Schwartz and colleagues [51] and Sehba and co-workers [52] published evidence that acute vasoconstriction contributes to ischemia, at least in part, through decreased NO availability. The authors showed that cerebral NO levels are acutely depressed after SAH, and that intra-arterial administration of the NO donor S-nitrosogluthathione ameliorates the LCBF changes and extra cellular glutamate release that occur in response to this event. However, the results of these studies do not explain why subarachnoid blood constricts the arteries in the subarachnoid space in delayed fashion, as has been demonstrated in experiments in which blood was placed or injected around blood vessels to induce vasospasm.

The most tenable and widely held hypothesis is that vasospasm is caused by contractile substances released from the blood clot, and/or by reactions initiated by the blood clot that generate contractile substances [38, 65]. Stoodley *et al.* [56] clearly demonstrated that vasospasm requires the presence of subarachnoid blood for at least 3 days after the initial haemorrhage. The removal of blood on or before 3 days post-SAH results in resolution of the initial vasospasm [25, 49, 56]. It has been postulated that haemoglobin released from the subarachnoid erythrocytes likely causes vasospasm [17, 39, 42].

In conclusion, our model provides significant improvements over previous techniques to produce SAH in rats. Unlike in models in which blood is placed in cisterna magna, basilar artery puncture results in extensive blood clot formation in the basal cisterns. Also because of the vascular injury the endothelium is exposed to subarachnoid space and potentially vasoactive substances are released into the cerebrospinal space. Also our model unlike in models which SAH produced with intraluminal suture allows controllable amount of haemorrhage, possibility of recurrent haemorrhages, topical application of vasodilatator agents, removal of clot or substrate collecting around the artery for reducing vasoconstriction and also opportunity to observe the arterial changes in a long periodical time (cranial window). And in our model there are no complications such as intracerebral haemorrhage or failure to produce SAH as reported in the literature.

References

1. Alkan T, Kahveci N (1999) Sıcanlarda anatomic, patolojik ve norolojik olarak vertebrobasilar arter okluzyon yontemlerinin arastırılması.Bursa Devlet Hastahanesi Bulteni 14: 153–158
2. Alkan T, Tureyen K, Ulutas M, Kahveci N, Goren B, Korfali E, Ozlük K (2001) Acute and delayed vasoconstriction after subarachnoid hemorrhage: local cerebral blood flow, histopathology, and morphology in the rat basilar artery. Arch Physiol Biochem 109: 145–153
3. Asano T, Sano K (1977) Pathogenetic role of no reflow phenomenon in experimental subarachnoid hemorrhage in dogs. J Neurosurg 46: 454–466
4. Baker KF, Zervas NT, Pile-Spellman J, Vacanti FX, Miller D (1987) Angiographic evidence of basilar artery constriction in the rabbit: a new model of vasospasm. Surg Neurol 27: 107–112
5. Barry KJ, Gogjian MA, Stein BM (1979) Small animal model for investigation of subarachnoid hemorrhage and cerebral vasospasm. Stroke 10: 538–541
6. Barth KN, Onesti ST, Krauss WE, Solomon RA (1992) A simple and reliable technique to monitor intracranial pressure in the rat: technical note. Neurosurg 30: 138–140
7. Bederson JB, Germano IM, Guarino L (1995) Cortical blood flow and cerebral perfusion pressure in a new noncraniotomy model of subarachnoid hemorrhage in the rat. Stroke 26: 1086–1091
8. Bederson JB, Levy AL, Ding WH, Kahn R, DiPerna CA, Jenkins AL, Vallabhajosyula P (1998) Acute vasoconstriction after subarachnoid hemorrhage. Neurosurg 42: 352–362
9. Bevan JA, Bevan RO, Frazee JG (1987) Functional arterial changes in chronic cerebrovasospasm in monkeys: an in vitro assessment of the contribution to arterial narrowing. Stroke 18: 472–481
10. Boisvert DPJ, Weir BKA, Overton TR, Reiffenstein RJ, Grace MGA (1979) Cerebrovascular responses to subarachnoid blood and serotonin in the monkey. J Neurosurg 50: 441–448
11. Brawley BW, Strandes DE, Kely WA (1968) The biphasic response of cerebral vasospasm in experimental subarachnoid hemorrhage. J Neurosurg 28: 1–8
12. Clower BR, Smith RR, Haining JL, Lockard J (1981) Constrictive endarteropathy following experimental subarachnoid hemorrhage. Stroke 12: 501–508
13. Delgado TJ, Arbab MA, Wgberg J, Svengaard NA (1988) The role of vasopressin in acute cerebral vasospasm: effect on spasm of vasopressin antagonist or vasopressin antiserum. J Neurosurg 68: 266–273
14. Delgado TJ, Brismar J, Svengaard NA (1985) Subarachnoid hemorrhage in the rat: angiography and fluorescence microscopy of the major cerebral arteries. Stroke 16: 595–602
15. Delgado TJ, NH, Svengaard NA (1986) Subarachnoid hemorrhage in the rat. Cerebral blood flow and glucose metabolism after selective lesions of the catecholamine systems in the brainstem. J Cereb Blood Flow Metab 6: 600–606
16. Delgado-Zygmunt TJ, Arbab MA, Shiokawa Y, Svendgaard NA (1992) A primate model for acute and late cerebral vasospasm: angiographic findings. Acta Neurochir (Wien) 118: 130–136
17. Duff TA, Louie J, Feilbach JA (1988) Erythrocytes are essential for development of cerebral vasculopathy resulting from subarachnoid hemorrhage in the cat. Stroke 19: 68–72
18. Echlin F (1971) Experimental vasospasm, acute and chronic, due to blood in the subarachnoid space. J Neurosurg 35: 646–656
19. Edwards DH, Byrne JV, Griffith TM (1992) The effect of chronic subarachnoid hemorrhage on basal endothelium-derived relaxing factor activity in intrathecal cerebral arteries. J Neurosurg 76: 830–837
20. Espinosa F, Weir B, Shnitka T (1986) Electron microscopy of simian cerebral arteries after subarachnoid hemorrhage and after the injection of horseradish peroxidase. Neurosurg 19: 935–945
21. Faraci FM, Brian JE (1994) Nitric oxide and the cerebral circulation. Stroke 25: 692–703
22. Findlay JM, Weir BK, Kanamaru K, Espinosa F (1989) Arterial wall changes in cerebral vasospasm. Neurosurg 25: 736–746
23. Fischer CM, Kistler JP, Davis JM (1980) Relation of cerebral vasospasm to subarachnoid hemorrhage visualized by computerized tomographic scanning. Neurosurg 6: 1–9
24. Fitch W, Pickard JD, Tamura A, Graham DI (1988) Effects of hypotension induced with sodium nitroprusside on the cerebral circulation before, and one week after, the subarachnoid injection of blood. J Neurol Neurosurg Psychiatry 51: 88–93
25. Handa Y, Weir BKA, Nosko M (1987) The effect of timing of clot removal on chronic vasospasm in a primate model. J Neurosurg 67: 558–564
26. Hashi K, Meyer JS, Shinmaru S, Welch KMA, Teraura T (1972) Haemodynamic and metabolic changes in experimental subarachnoid hemorrhage in monkeys. Eur Neurol 8: 32–37
27. Iadecola C (1993) Regulation of the cerebral microcirculation

during neural activity: is nitric oxide the missing link? Trends Neurosci 16: 206–214
28. Jackowski A, Crockard A, Burnstock G, Russel RR, Kristek F (1990) The time course of intracranial pathophysiological changes following experimental subarachnoid hemorrhage in the rat. J Cereb Blood Flow Met 10: 835–849
29. Jakubowski J, Bell BA, Symon L, Zawirski MB, Francis DM (1982) A primate model of subarachnoid hemorrhage: changes in regional cerebral blood flow, autoregulation, carbon dioxide reactivity, and central conduction time. Stroke 13: 601–611
30. Jakubowski J, McCleery WNC, Todd JH, Smart RC (1976) Cerebral blood flow changes in acute experimental hmorrhagic vasospasm. Acta Neurochir (Wien) 34: 265–268
31. Kader A, Krauss WE, Onesti ST, Elliot JP, Solomon RA (1990) Chronic cerebral blood flow changes following experimental subarachnoid hemorrhage in rats. Stroke 21: 577–581
32. Kajita Y, Suzuki Y, Oyama H, Tanazawa T, Takayashu M, Shibuya M, Sugita K (1994) Combined effect of L-arginine and superoxide dismutase on the spastic basilar artery after subarachnoid hemorrhage in dogs. J Neurosurgery 80: 476–483
33. Kaplan B, Brint S, Tanabe J, Jacewics M, Wang X-J, Pulsinelli W (1991) Temporal thresholds for neocartical infarction in rats subjected to reversible focal cerebral ischemia. Stroke 22: 1032–1039
34. Kassell NF, Haley EC Jr, Apperson-Hansen C, Alves WM (1996) Randomized, double-blind, vehicle-controlled trial of tirilazad mesylate in patients with aneurysmal subarachnoid hemorrhage: a cooperative study in Europe, Australia, and New Zealand. J Neurosurg 84: 221–228
35. Kuyama H, Ladds A, Branston NM, Nitta M, Symon L (1984) An experimental study of acute subarachnoid hemorrhage in baboons: changes in cerebral blood flow, electrical activity and water content. J Neurol Neurosurg Psychiatry 47: 354–364
36. Liszczak TM, Black PM, Tzouras A, Foley L, Zervas NT (1984) Morphological changes of the basilar artery, ventricles, and choroid plexus after experimental subarachnoid hemorrhage. J Neurosurg 61: 486–493
37. Liszczak TM, Varsos VG, Black PM, Kistler JP, Zervas NT (1983) Cerebral arterial constriction after experimental subarachnoid hemorrhage is associated with blood components within the arterial wall. J Neurosurg 58: 18–26
38. MacDonald RL (1995) Cerebral vasospasm. Neurosurgery Q 5: 73–97
39. Macdonald RL, Weir BK (1991) A review of hemoglobin and the pathogenesis of cerebral spasm. Stroke 22: 971–982
40. Macdonald RL, Weir BK, Grace MG, Martin TP, Doi M, Cook DA (1991) Morphometric analysis of monkey cerebral arteries exposed in vivo to whole blood, oxyhemoglobin, methemoglobin, and bilirubin. Blood Vessels 28: 498–510
41. Macdonald RL, Weir BK, Runzer TD, Grace MG, Findlay JM, Saito K, Cook DA, Mielke BW, Kanamaru K (1991) Etiology of cerebral vasospasm in primates. J Neurosurg 45: 415–424
42. Mayberg MR, Okada T, Bark DH (1990) The role of hemoglobin in arterial narrowing after subarachnoid hemorrhage. J Neurosurg 72: 634–640
43. Megyesi JF, Vollrath B, Cook DA, Findlay JM (2000) In vivo animal models of cerebral vasospasm. Neurosurg 46: 448–461
44. Megyesi JF, Findlay JM, Vollrath B, Cook DA, Chen MH (1997) In vivo angioplasty prevents the development of vasospasm in canine carotid arteries: pharmacological and morphological analyses. Stroke 28: 1216–1224
45. Megyesi JF, Vollrath B, Cook DA, Chen MH, Findlay JM (1999) Long-term effects of in vivo angioplasty in normal and vasospastic canine carotid arteries: pharmacological and morphological analyses. J Neurosurg 91: 100–108
46. Mendelow AD, McCalden TA, Hattingh J, Coull A, Rosendorff C, Eidelman BH. (1981) Cerebrovascular reactivity and metabolism after subarachnoid hemorrhage in baboons. Stroke 12: 58–65
47. Nakagomi T, Kassell NF, Sasaki T, Fujiwara S, Lehman RM, Torner JC (1987) Impairment of endothelium-dependent vasodilation induced by acetylcholine and adenosine triphosphate following experimental subarachnoid hemorrhage. Stroke 18: 482–489
48. Naveri L, Stromberg C, Saavedra JM (1994) Angiotensin IV reverses the acute cerebral blood flow reduction after experimental subarachnoid hemorrhage in the rat. J Cereb Blood Flow Metab 14: 1096–1099
49. Nosko M, Weir BKA, Lunt A (1987) Effect of clot removal at 24 hours on chronic vasos pasm after SAH in the primate model. J Neurosurg 6: 416–422
50. Petruk KC, West GR, Marriott MR, McIntyre JW, Overton TR, Weir BK (1972) Cerebral blood flow following induced subarachnoid hemorrhage in the monkey. J Neurosurg 37: 316–324
51. Schwartz AY, Sehba F, Bederson J (2000) Decreased nitric oxide availability contributes to acute cerebral ischemia after subarachnoid hemorrhage. Neurosurg 47: 208–215
52. Sehba F, Ding WH, Chereshnev I, Bederson JB (1999) Effects of S-nitrosoglutathione on acute vasoconstriction and glutamate release after subarachnoid hemorrhage. Stroke 30: 1955–1961
53. Seifert V, Stolke D, Reale E (1989) Ultrastructural changes of the basilar artery following experimental subarachnoid hemorrhage. A morphological study on the pathogenesis of delayed cerebral vasospasm. Acta Neurochir (Wien) 100: 164–171
54. Sobey CG, Faraci FM (1998) Subarachnoid haemorrhage: what happens to the cerebral arteries? Clin Exp Pharmacol Physiology 25: 867–876
55. Solomon RA, Antunes JL, Chen RY, Bland L, Chien S (1985) Decrease in cerebral blood flow in rats after experimental subarachnoid hemorrhage: a new animal model. Stroke 16: 58–64
56. Stoodley M, Weihl CC, Zhang Z, Lin G, Johns LM, Kowalczuk BA, Ghadge G, Roos RP, Macdonald RL (2000) Effect of adenovirus-mediated nitric oxide synthase gene transfer on vasospasm after experimental subarachnoid hemorrhage. J Neurosurg 93: 463–470
57. Swift DM and Solomon RA (1988) Subarachnoid hemorrhage fails to produce vasculopathy or chronic flow changes in rats. Stroke 19: 878–882
58. Taneba Y, Sakata K, Yamada H, Ito T, Takoda Y (1978) Cerebral vasospasm and ultrastructural changes in the arterial wall. An experimental study. J Neurosurg 49: 229–239
59. Toda N (1988) Hemolysate inhibits cerebral artery relaxation. J Cereb Blood Flow Metab 8: 46–53
60. Toda N, Ayajiki K, Okamura T (1993) Endothelial modulation of contractions caused by oxyhemoglobin and NG-nitro-L-arginine in isolated dog and monkey cerebral arteries. Stroke 24: 1584–1589
61. Toda N, Shimizu K, Ohta T (1980) Mechanism of cerebral arterial contraction induced by blood constituents. J Neurosurg 53: 312–322
62. Trojanowski T (1982) Experimental subarachnoid hemorrhage, I: a new approach to subarachnoid haemorrhage in cats. Acta Neurochir (Wien) 62: 171–175
63. Umansky F, Kaspi T, Shalit MN (1983) Regional cerebral blood flow in the acute stage of experimentally induced subarachnoid hemorrhage. J Neurosurg 58: 210–216
64. Veelken JA, Laing RJC, Jakubowski J (1995) The Sheffield model of subarachnoid hemorrhage in rats. Stroke 26: 1279–1284

65. Weir B, MacDonald RL (1993) Cerebral vasospasm. Clin Neurosurg 40: 40–55
66. Yamaura I, Tani E, Mada Y, Inami N, Shindo H (1992) Endothelin-I in canine artery in vasospasm. J Neurosurg 76: 99–105
67. Yokoto M, Tani E, Maeda Y, Kokubu K (1987) Effects of 5-Lipoxygenase inhibitor on experimental delayed cerebral vasospasm. Stroke 18: 512–518
68. Zea Longa E, Weunstein PR, Carlson S, Cummins R (1992) Reversible middle cerebral artery occlusion without craniectomy in rats: Stroke 20: 84–91

Correspondence: E. Korfali, M.D., Uludag University, School of Medicine, Department of Neurosurgery, 16059 Bursa, Turkey.

Acta Neurochir (2002) [Suppl] 83: 71–78

Methods of Developmental Research

A. Ünlü

Neurosurgery Department, University of Ankara, School of Medicine, Ankara, Turkey

Summary

Neural tube defects (NTD) caused by abnormal neurulation are the major congenital anomalies which result in fetal or embryonic death, and medical, financial and social problems. The multifactorial events of neurulation have attracted researchers to identify the mechanisms of this disability.

Research focused on NTDs is one of the major topics in developmental experiments. Mammalian, avian, amphibian and computer models are used as fundamental models to discover specific events causing NTDs. There are advantages of working on some models: rats and mice are mammalian models of neurulation; amphibians and avian embryos are simple models and more practical.

Advancement in laboratory techniques has yielded more detailed information about neurulation which will assist in future with prevention and therapy of these defects.

Keywords: Method; neural tube defect; neurulation; avian embryo.

Introduction

The neural system has long been the subject of interest for many researchers with stroke, trauma, tumors, congenital anomalies, degenerative processes, and regeneration being the major topics of the experiments. While some of the experiments have been focused on the search for the developmental basis of problems, others have explored clues gathered from developmental models. Neural tube defects (NTD) are major subject of developmental research.

Neural tube defects, resulting in fetal or embryonic death and functional disabilities, are causing medical, financial and social problems. To prevent or reduce their effects, the underlying mechanism must be clarified. The developing stages of the nervous system should first be thoroughly studied [5, 24, 30, 31].

Neurulation has been defined as formation of the neural tube. In the primary neurulation period, the cranial and most of the spinal levels are formed. Development of the most caudal part of the spinal cord from a pluripotent cell mass has been named secondary neurulation. Neurulation occurs simultaneously with the development of other parts of the embryo.

Neural induction, *formation of the neural plate*, *shaping of the neural plate*, *bending*, *folding*, and *fusion of folds* are the main steps of neurulation. Secondary neurulation, which accompanies the final events, precedes the closure of the neuropores. Thus, the *neural tube* is formed.

In gastrulation and early neurulation periods, cells in the epiblast layer, which is one of the embryonic layers such as the hypoblast and mesoderm, show some differentiation. Cells change their columnar structure to pseudostratified high columnar type. The thickness of the layer is larger around the midline than it is laterally. Some of the epiblast cells migrate between the epiblast and hypoblast, and form head process that will later form the notochord and gut endoderm. The notochord, sunk in the hypoblast, changes its position between the endoderm and ectoderm and forms a tubular structure. The notochord constitutes an axis in certain embryos; however, being thinner in mammalians, it does not form a skeletal structure.

Neural induction is promoted by some molecules originating from the node (chordin, follistation, noggin), notochord, endoderm and mesoderm. Node-originated molecules inhibit the bone morphogenic protein (BMP) expressed by the epiblast, and change some of the cells into neuroepithelial type, which then form the neural plate. The future cranial part is wide, while the spinal part is narrow. Throughout the subsequent stages the shape becomes a more elongated structure.

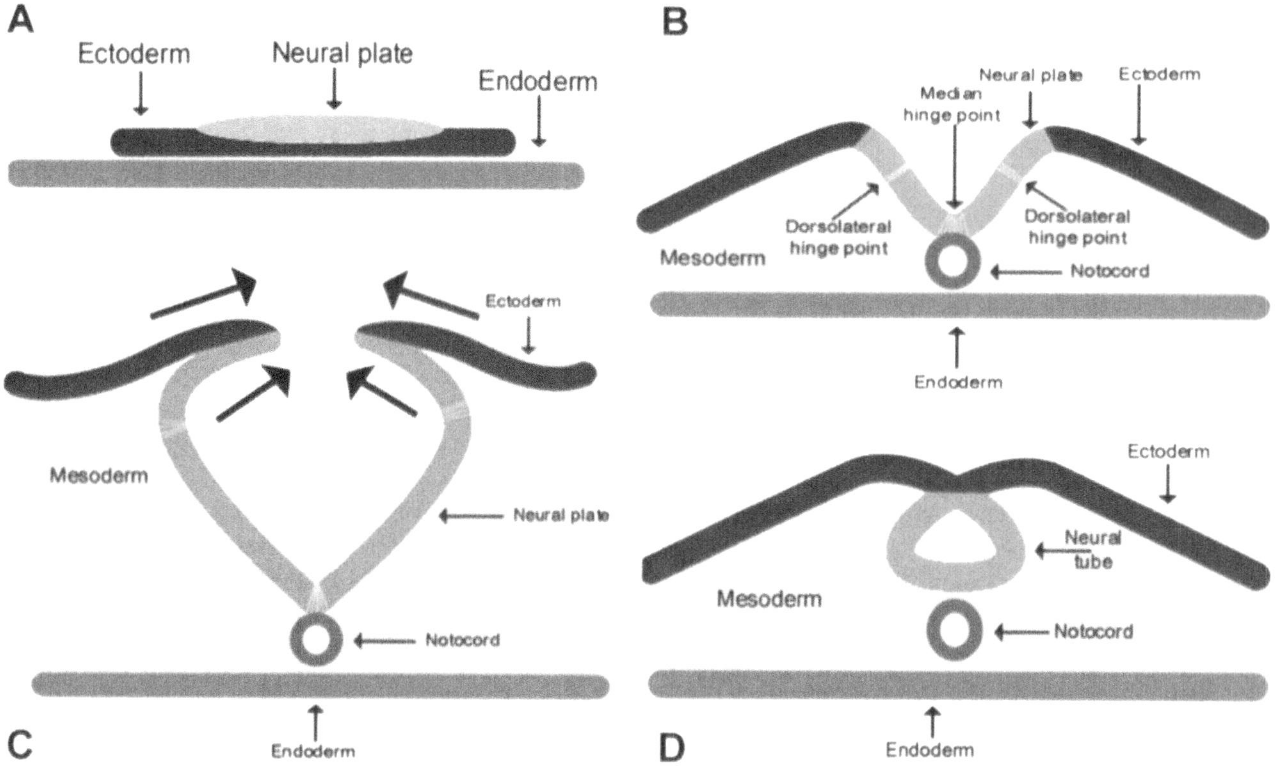

Fig. 1. Drawings of fundamental neurulation steps: formation of neural plate, ectoderm and endoderm (A); furrowing and folding of neural plate, dorsolateral hinge points, median hinge points (B); convergence of neural plate to dorsal midline (C); fusion of neural plate and formation of neural tube; ectoderm also fuses at dorsal midline (D)

Shaping of the neural plate is a result of some cellular events. There is widening of the neural plate at future brain regions and narrowing at future spinal regions. Rearrangements in cellular morphology and cell division are possible factors for these changes in neural plate shape. Actually, neuroepithelial cells are spindle shaped and have different characteristics related to rostrocaudal and mediolateral orientation. Unlike in the mediolateral axis, elongation of cell is a unique event throughout the rostrocaudal axis. While cells are taller in lateral regions, the shorter cells in the midline, which are anchored and affected by the notochord (with Sonic hedgehog gene), are defined as median hinge points (MHP). There is another region in the lateral plate containing similar cells termed dorsolateral hinge points (DLHP). The orientation and polymerization of microtubules in cells may be responsible for elongation. Since the volume of the cell is constant, mediolateral narrowing is inevitable. Cell cycle and position of daughter cells are other key factors in neural plate shaping. Daughter cells move in the rostrocaudal axis in spinal regions and in both mediolateral and rostrocaudal axes in brain regions. Cell cycle time is also different in the mediolateral axis. While lateral cells (L) have shorter cell cycle time, MHP cells have longer cell cycle time (Fig. 1).

While shaping events occur, some changes related with bending of the neural plate also take place. As mentioned before, cells change their morphology in the shaping period. The resultant form is spindle in shape, but this is not unique to all cells. Depending on the nucleus position, cells could be bulbous or wedge shaped. In MHP, the majority of the cells are wedge shaped with their nucleus positioned at the base near the basal membrane. Wedging of these cells is necessary for furrowing. Like MHP cells, DLHP cells that have been anchored to the ectoderm have similar morphologic changes, changing to wedge shape. As shown in amphibians, median wedging is not essential for elevation and convergence, and is only needed for furrowing. But wedging of DLHPs has a crucial role in bending. This supports the idea that bending is a multifactorial event. Some external factors acting on bending have been suggested. Non-neural ectodermal cells have similar morphological changes during bending. Cell division and movement of the cells are also important factors like in neuroepithelial cells. After division, cells move at the rostrocaudal and lateromedial axes. The higher cell proliferation and tight cellular interactions expand the surface ectoderm at the mediolateral axis. Similar events occur in deeper layers like the mesoderm and endoderm. Growth of the embryo and notochord in the rostrocaudal axis facilitates stretching of the neural plate which has been shown to affect bending (Fig. 1).

External forces are assumed to affect folding. The surface ectoderm especially facilitates folding and convergence, the result of which is the tip of the neural plate meeting at the midline to form the roof plate of the neural tube. While moving to midline, the extracellular matrix plays a role in separating surface ectoderm and the neural plate. The extracellular matrix is later filled with migrated neural crest cells. The surface ectoderm also meets at the midline over the neural tube (Fig. 1).

Fusion of the folds is the last step of tube formation. Digitations in cell structure and cell surface molecules (glycoconjugates) are key features of fusion. Fusion begins in the spinal cord hindbrain border. Similar closure sites are situated at the forebrain (anterior neuropores) and caudal spine (posterior neuropores). Like a zipper, fusion of the folds progresses in cranial and caudal directions and completes the form of the neural tube (Fig. 1).

Secondary neurulation is the step to forming the most caudal end of the spinal cord. The morphologic changes of tail bud cells and cellular organization are the main events of secondary neurulation [28].

Materials and Methods

Although information about human embryonic development has been gathered [24], details of the neurulation mechanisms are unclear. The main target of developmental experiments is clarification of human neurulation, so the research model should also match the human model. Unfortunately, there is no homologous model for humans. Monkeys have been used previously despite certain limitations like a longer incubation period. The rat [22], mouse [19], amphibian (Xenopus laevis) [18], avian (chicken, quail) [26], rabbit [25], zebrafish [15] and computer models [9] are the other models used in developmental research.

Since neurulation is a time-dependent event, it is essential to identify the steps of neurulation. Stages differ between species. While stage 9 corresponds to day 20 in humans, it is day 1 in the chicken and day 8 in the rabbit. The staging system used in experiments should match the animal model. Butler and Juurlink [3] made a detailed comparison of the stages. The Hamburger-Hamilton staging system has been widely used for staging chicken embryos [16].

Avian Model

Egg

Eggs can be obtained from manufacturers. The White Leghorn type is most commonly used in embryologic experiments. The eggs should be transported below 21–22 °C. Embryonic development begins over 25–26 °C in fertilized eggs. The eggs could be stored a few days below 22 °C with 70–80% relative humidity. The sharper end should be down. There should be no rapid temperature changes that could affect the embryo.

Laboratory

All procedures should be done under sterile conditions. Incubators, glass containers, watch-glass, glass rings, microinstruments, Hamilton microsyringes, and dissection microscope are the basic tools used for harvesting embryos.

Incubation

Fertilized eggs are incubated at 38–40 °C, in 95% humidity. A rotator is mandatory to move the eggs, to prevent embryos from sticking into the inner side of the shell. The development of the embryo stops below 27 °C.

Procedures

Two methods have been used for embryo development procedures. Fineman and Schoenwolf described the "*windowing method*" with some limitations [11]. The method was revised in order to prevent embryonic distress that caused dysmorphogenesis (NTD and early amnion deficit spectrum) [13]. The other technique suitable for in vitro cultures was described by New [23].

Windowing Method

In the modified technique, the shell was opened from the narrow end, after which the empty space was filled with albumen or saline. The shell was sealed with plastic tape and rotated upside down (Fig. 2A–D).

In this technique, some teratogens have been used for subblastodermic injections and embryos cultured in vivo. Hamilton microsyringes have been used for

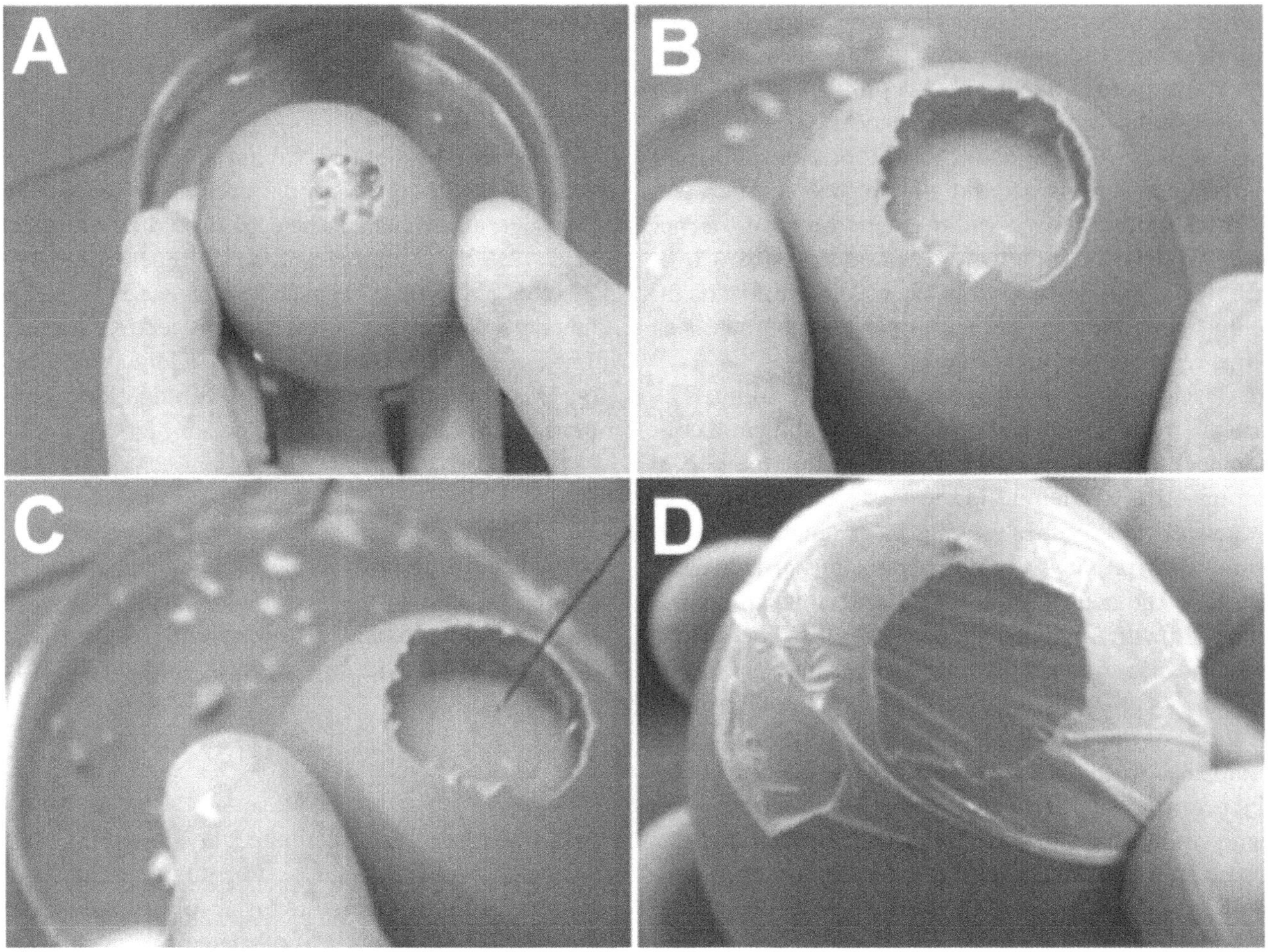

Fig. 2. Windowing method: opening of egg shell (A), widening of the shell and embryo (B), sublastodermic injection (C) and drape sealed egg shell (D)

injections; 20 μl of solutions are sufficient to protect the embryo from shearing forces (Fig. 2C).

New's Technique

New's method has been widely used for in vitro cultures.

A glass container with sterile saline is necessary. The depth of the water should be 3–5 cm. The temperature of the saline must be 38 °C and all procedures should be carried out at that temperature within the container. A Petri dish, watch-glass, and glass ring with a diameter of 2 cm should be placed in the container.

The shell is opened from the wide end. A 2 × 2 cm opening is needed to pour the albumen outside (Fig. 3A). The yolk should be carefully protected from shearing by the cut ends of the shell. The albumen is kept in a Petri dish that may be needed later for a nutrient preparation. The ligamentous attachment of albumen to yolk should be cut with a microscissors (Fig. 3B). The opening is widened and the shell that contains yolk and embryo is put in the glass container (Fig. 3C, D). The yolk is then transferred into the water and cut at the equator while the embryonic side is up (Fig. 3E). The vitelline membrane containing the embryo is freed from the yolk and moved to the watch-glass with fine forceps (Fig. 3F, G). A glass ring is put over the membrane that centers the embryo (Fig. 3H).

The watch-glass is moved outside the container and the water is emptied. The fine adjustment of water is done with a piece of cotton. The membrane and embryo are inflated with a nutrient medium; deep wrinkles should be prevented. The watch-glass containing the embryo is put in a Petri dish and transferred into an incubator.

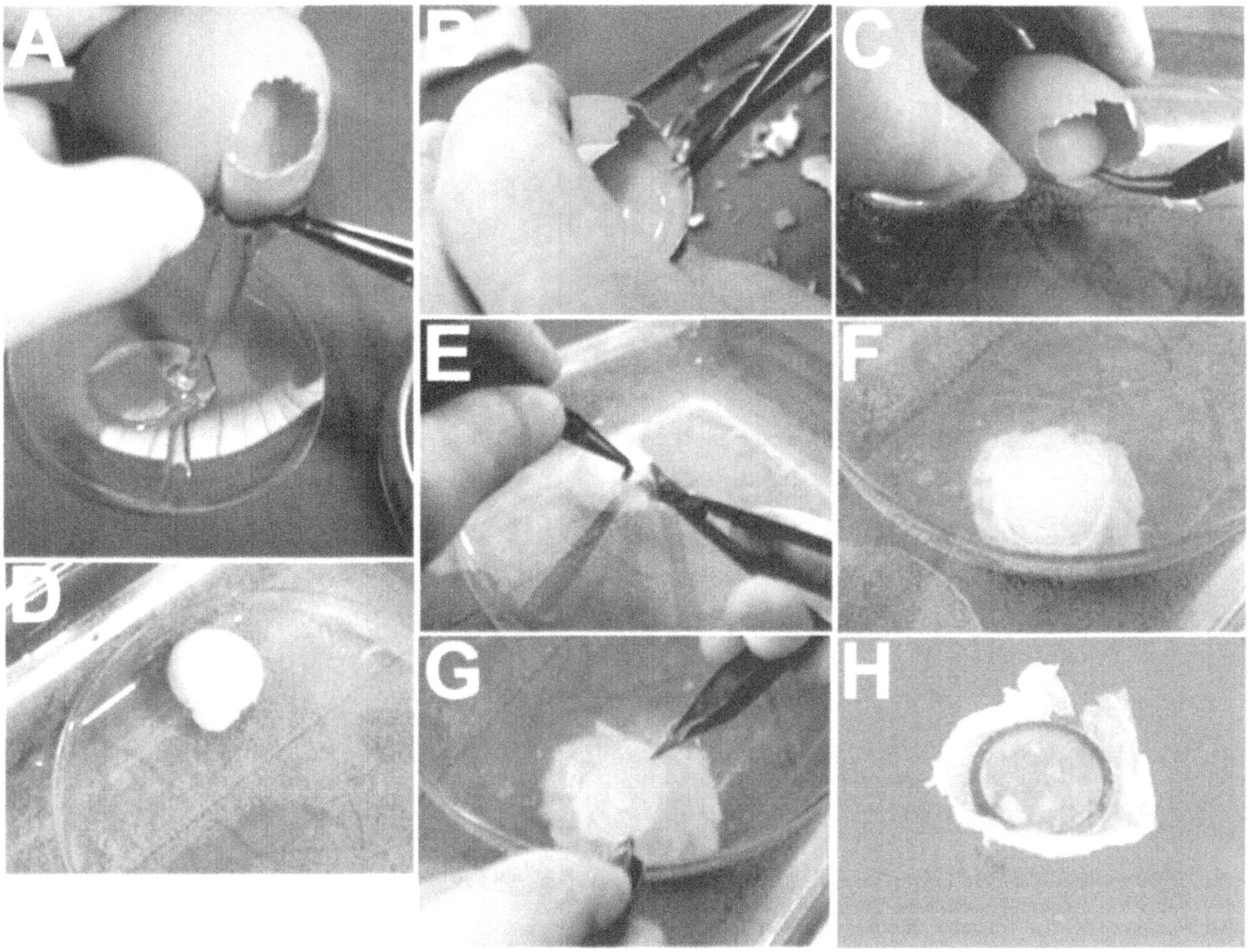

Fig. 3. New method for embryo explantation: removal of albumen (A); widening of the shell with forceps (B); transportation into water (C); cutting of the yolk with microscissor at equator (D,E); transportation of the yolk-free vitelline membrane that contains embryo to watch-glass in water (F,G); covering of the membrane with glass ring (H)

Nutrients

The thin albumen harvested from the egg has been commonly used as a nutrient. Sterile saline containing antibiotics at pH 7.4 could also be used (Coon solution: F12 solution + 100 IU penicillin + 1 µg/ml streptomycin). For shorter incubation only saline could be used.

Histological Preparation

There is not much difference in preparation from other techniques. Paraffin techniques, plastic embedding, and whole mounting are basic methods (Fig. 4). For fluorescent or electron microscopic images, specific techniques should be used.

Polymerase chain reaction (PCR), in situ hybridization, and blotting techniques are necessary for identifying protein and gene expressions.

Discussion

Some properties of the animal model contribute to the choice of the perfect model to fit the hypothesis. Availability, surgical manipulation, incubation period, genetic properties, and specific properties related with neurulation itself may affect model choice.

Mammalian models are difficult to conduct research on due to their complex structures, although they are suitable for genetic-based research. The neurulation procedure is also complicated due to other embryonic developmental events when compared to others. But

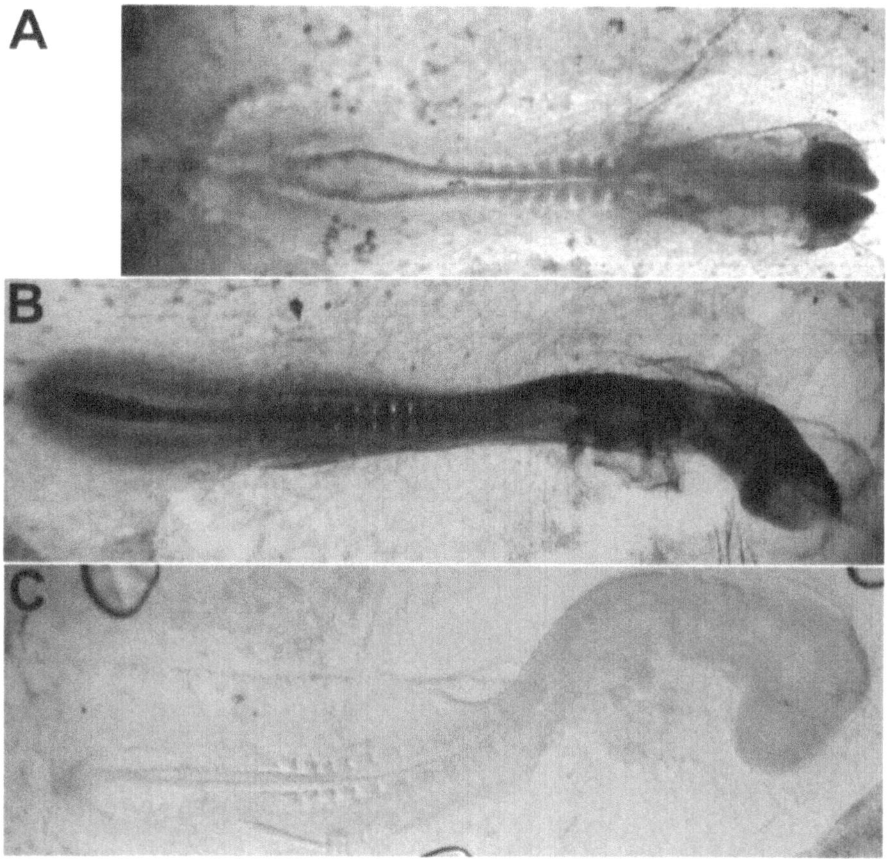

Fig. 4. Whole mounted chicken embryos: Stage 9 embryo stained with HE. Posterior and anterior neuropores are open (A). Stage 13 embryo stained with HE. Anterior neurupore is closed. Cranial and cervical flexures are visible (B). Stage 14 embryo without staining. Right-angled cranial flexure and heart at ventral of the embryo are visible (C)

some of the genetic properties of rats and mice resemble human genetics. Fly and nematodes, which have simple and genetic properties similar to humans, have also been used for genetic characteristics of neurulation. Mutant embryos, like Loop-tail (Lp/Lp) embryos (slow axial elongation rate, abnormal ECM composition), Splotch (Sp/Sp) and delayed Splotch (Sp^d/Sp^d) embryos (neural crest defect), curly tail (ct/ct) embryos (extraneural cell proliferation defect, increased ventral curve), and trisomy 12 and 14 (mesodermal defect) embryos have been used for genetic-based experiments [6].

Amphibian embryos are easy to handle and suitable for cell tracing experiments. Cells are pigmented and larger than others. Embryos are accessible for surgical manipulations like implantation or explantation. Limited cellular division without accompanying morphologic changes is another advantage of amphibians making neurulation simpler than in others [17].

Chicken embryos are also suitable for surgery. They have patterns of neurulation similar to humans at spinal levels, and the incubation period is shorter. Direct approach and in vivo or in vitro cultures are advantages of avian embryos, which have been used in most of the teratology and NTD researches [27].

In rabbit embryos, the axial curvature, which has been shown to influence neurulation, is absent at certain stages. It was concluded that, as in chicken embryos, the closure rates are higher since there is a stage without curvature in whole embryonic morphology [25].

Culturing of the embryo is the next step in neurulation experiments. In vivo, in ovo, and in vitro culturing have been widely used. One of the purposes is to observe the progress of neurulation. In vitro cultures are simple and facilitate easy monitoring of the steps of neurulation. Videos of embryonic development have been recorded successfully (http://anatomy.med.

unsw.edu.au/cbl/embryo/Movies/Movies.htm). Tissue transplantations like notochord or ectoderm could be done under microscope on cultured embryos. The advantage of transplantation is to identify the effects or relations of those tissues.

Histological examination can be done by light microscopy, fluorescent microscopy, confocal microscopy and electron microscopy (EM) (Scanning Electron Microscopy, Transmission Electron Microscopy). EM is preferable for cellular details. Tissue preparation protocols are not different from other tissues. Cell shape differences, intracellular organizations, cellular movements, and proliferation can be displayed by microscopic examinations. Ferreira *et al.* used phallicidin, which is labeled by fluorescent tracer to determine the organization of intracellular filaments [10]. Confocal and electron microscopy were used in both living and fixed specimens of the chicken embryo. The advantage of confocal microscopy is that it demonstrates cellular changes in the living embryo. Whole mounting or sections parallel or perpendicular to the embryonic axis could be used in combination with molecular biology protocols (in situ hybridization, blotting, immunocytochemistry, etc.) [21]. Fate mapping and fluorescent tracing methods reveal information about cellular movements, one of the most important events of neurulation. Schoenwolf used a fluorescent-histochemical marker (peroxidase-rhodamine isothiocyanate – R-HRP) and perpendicular sections to show the tissue movements in the epiblast layer. Major displacements and relations with the node had been documented in shaping and bending of the neural plate [29]. In the experiment of Dale *et al.*, fate mapping with fluorescent tracing was combined with in situ hybridization and immunocytochemistry [8]. The combination of these techniques revealed information about cell migration and cell-to-cell interactions.

Molecular interactions occur simultaneously with cellular morphologic changes in neurulation. At every step, significant regulatory induction or inhibition takes place at different localizations and times. Protein, RNA, DNA and gene expressions are responsible for cell-to-cell interactions. Non-neural and neural (notochord, endoderm, ectoderm and extracellular matrix) communication determines the fate of the neurulation. Additional methods should be added to principle methods in order to identify relations. Blotting techniques, polymerase chain reaction (PCR, DD-PCR), in situ hybridization, and chemical or immunocytochemical techniques are necessary to discover those contacts [35]. Gene expression patterns and effects of the genes are other important points of interest in neurulation. Blottting techniques could be used to search specific DNA fragments. Wheeler *et al.* used inducible gene expression technique, which is a more complicated protocol, to identify some specific genes and effects in development of the embryo [33]. Advancement in gene technology led to usage of microarray and gene chip techniques, which might facilitate discovery of the gene sequences at the basis of neurulation [20, 32, 34].

Another method is the computer model of neurulation. Computer programs have simulated morphological changes and cellular movements. The intrinsic and extrinsic forces that drive neurulation have also been simulated. Microtubules and microfilaments (which affect cellular morphology), and effects of the notochord were shown in this computer model [4]. In another computer model used by Bush, the effect of apical constriction in bending of the neural plate was demonstrated [2]. The disadvantage of this model is inadequate simulation of all factors acting on neurulation. Nevertheless, it should be accepted that significant mechanical information in cell morphology has been gathered with this technique.

Another aspect of the experiments is to demonstrate the relation between genetic background and environmental effects that cause NTDs [12]. Breen [1] and Friedberg [14] documented hyperthermia and radiation-induced NTDs. Many toxicology and teratology experiments have also been conducted to show the side effects on neural development. A summary of the experiments was well documented in a table by Copp *et al.* [7]. The experiments were designed to demonstrate the effects of various agents given maternally or in in vitro cultures.

There is no certain model suitable for neurulation experiments since it is the result of multifactorial events. It is essential to use multiple methods which will provide sufficient clues about neurulation. In design of developmental experiments, every step of neurulation should be carefully reviewed, and a hypothesis should be demonstrated with appropriate models.

References

1. Breen JG, Claggett TW, Kimmel GL, Kimmel CA (1999) Heat shock during rat embryo development in vitro results in decreased mitosis and abundant cell death. Reprod Toxicol 13(1): 31–39

2. Bush KT, Lynch FJ, DeNittis AS, Steinberg AB, Lee HY, Nagele RG (1990) Neural tube formation in the mouse: a morphometric and computerized three-dimensional reconstruction study of the relationship between apical constriction of neuroepithelial cells and the shape of the neuroepithelium. Anat Embryol 181(1): 49–58
3. Butler H, Juurlink BHJ (1987) An atlas for staging mammalian and chick embryos. CRC Press. Inc, Florida
4. Clausi DA, Brodland GW (1993) Mechanical evaluation of theories of neurulation using computer simulations. Development 118: 1013–1023
5. Copp A, Cogram P, Fleming A, Gerrelli D, Henderson D, Hynes A, Kolatsi-Joannou M, Murdoch J *et al* (2000) Neurulation and neural tube closure defects. Methods Mol Biol 136: 135–160
6. Copp AJ (1994) Genetic models of mammalian neural tube defects. Ciba Found Symp 181: 118–134
7. Copp AJ, Brook FA, Estibeiro JP, Shum AS, Cockroft DL (1990) The embryonic development of mammalian neural tube defects. Prog Neurobiol 35(5): 363–403
8. Dale K, Sattar N, Heemskerk J, Clarke JD, Placzek M, Dodd J (1999) Differential patterning of ventral midline cells by axial mesoderm is regulated by BMP7 and chordin. Development 126(2): 397–408
9. Dunnett D, Goodbody A, Stanisstreet M (1991) Computer modelling of neural tube defects. Acta Biotheor 39(1): 63–79
10. Ferreira MC, Hilfer SR (1993) Calcium regulation of neural fold formation: visualization of the actin cytoskeleton in living chick embryos. Dev Biol 159(2): 427–440
11. Fineman RM, Schoenwolf GC (1987) Animal model: dysmorphogenesis and death in a chicken embryo model. Am J Med Genet 27(3): 543–552
12. Finnell RH, Gelineau-van Waes J, Bennett GD, Barber RC, Wlodarczyk B, Shaw GM, Lammer EJ, Piedrahita JA *et al* (2000) Genetic basis of susceptibility to environmentally induced neural tube defects. Ann NY Acad Sci 919: 261–277
13. Fisher M, Schoenwolf GC (1983) The use of early chick embryos in experimental embryology and teratology: improvements in standard procedures. Teratology 27(1): 65–72
14. Friedberg W, Faulkner DN, Neas BR, Hanneman GD, Darden EB Jr, Deal RB Jr, Parker DE (1987) Dose-incidence relationships for exencephalia, anophthalmia and prenatal mortality in mouse embryos irradiated with fission neutrons or 250 kV X-rays. Int J Radiat Biol Relat Stud Phys Chem Med 52(2): 223–236
15. Halpern ME, Hatta K, Amacher SL, Talbot WS, Yan YL, Thisse B, Thisse C, Postlethwait JH *et al* (1997) Genetic interactions in zebrafish midline development. Dev Biol 187(2): 154–170
16. Hamburger V, Hamilton HL (1992) A series of normal stages in the development of the chick embryo. 1951 [classical article] [see comments]. Dev Dyn 195(4): 231–272
17. Jacobson AG (1994) Normal neurulation in amphibians. Ciba Found Symp 181: 6–21
18. Jacobson AG, Oster GF, Odell GM, Cheng LY (1986) Neurulation and the cortical tractor model for epithelial folding. J Embryol Exp Morphol 96: 19–49
19. Juriloff DM, Harris MJ (2000) Mouse models for neural tube closure defects. Hum Mol Genet 9(6): 993–1000
20. Lee PS, Lee KH (2000) Genomic analysis. Curr Opin Biotechnol 11(2): 171–175
21. Luque JM, Morante-Oria J, Riederer BM, Fairen A (2001) Whole-mount confocal immunofluorescence of mammalian CNS. Brain Res Prot 6: 129–133
22. Morriss-Kay G, Tuckett F (1985) The role of microfilaments in cranial neurulation in rat embryos: effects of short-term exposure to cytochalasin D. J Embryol Exp Morphol 88: 333–348
23. New DAT (1955) A new technique for the cultivation of the chick embryo in vitro. J Embryol Exp Morphol 3: 326–331
24. O'Rahilly R, Muller F (1994) Neurulation in the normal human embryo. Ciba Found Symp 181: 70–82
25. Peeters MC, Viebahn C, Hekking JW, van Straaten HW (1998) Neurulation in the rabbit embryo. Anat Embryol (Berl) 197(3): 167–175
26. Schoenwolf GC (1991) Cell movements driving neurulation in avian embryos. Development [Suppl] 2: 157–168
27. Schoenwolf GC (1994) Formation and patterning of the avian neuraxis: one dozen hypotheses. Ciba Found Symp 181: 25–38
28. Schoenwolf GC (1984) Histological and ultrastructural studies of secondary neurulation in mouse embryos. Am J Anat 169(4): 361–376
29. Schoenwolf GC (1985) Shaping and bending of the avian neuroepithelium: morphometric analyses. Dev Biol 109(1): 127–139
30. Schoenwolf GC, Smith JL (2000) Mechanisms of neurulation. Meth Mol Biol 136: 125–134
31. Smith JL, Schoenwolf GC (1997) Neurulation: coming to closure. Trends Neurosci 20(11): 510–517
32. Walter G, Bussow K, Cahill D, Lueking A, Lehrach H (2000) Protein arrays for gene expression and molecular interaction screening. Curr Opin Microbiol 3(3): 298–302
33. Wheeler GN, Hamilton FS, Hoppler S (2000) Inducible gene expression in transgenic Xenopus embryos. Curr Biol 10(14): 849–852
34. Wilgenbus KK, Lichter P (1999) DNA chip technology ante portas. J Mol Med 77(11): 761–768
35. Zhong GY, Riov J, Goren R, Holland D (2000) Competitive hybridization: theory and application in isolation and quantification of differentially regulated genes. Anal Biochem 282(1): 129–135

Correspondence: Agahan Ünlü, Tusso Bloklari, N-1 Blok No 14, 10. Cadde, Emek, 06510 Ankara, Turkey.

Acta Neurochir (2002) [Suppl] 83: 79–83

Model Systems in Neurooncology

J.-C. Tonn

Department of Neurosurgery, Ludwig-Maximilians-University, Munich, Germany

Summary

Cell- and molecular biological techniques have had a major impact on experimental neurooncology in recent years; yet we are lacking suitable model systems. Monolayer cell cultures are rapid, reproducible and reliable systems, however, their validity is of major concern. Three dimensional culture systems, especially derived from primary biopsies, match better with the in vivo situation albeit being more tricky to handle.

Animal models for glioma have to be orthotopic in order to draw any conclusions; most cell lines implanted into rodents still do not show the typical invasive phenotype. In addition, immunological phenomena have to be taken into account as well as changes of the biological features once cells have undergone the process of any transfection.

Keywords: Cell culture; spheroid systems; neurooncology; orthotopic glioma model.

Introduction

Preclinical and clinical research in neurooncology need tumour models. In basic science they serve as systems to characterize the proof of principle in antineoplastic therapy as well as the aim to decipher cell and molecular biological features of the tumour cells. Once a therapeutic concept has proven to be effective in vitro, in vivo models are needed to evaluate the concept prior to any launch of clinical study. On the way from bench to bedside, the first step in a clinical trial has to outline the toxicity of any new therapeutical concept, thereafter the proof of efficacy. Phase I and Phase II trials are designed in such a way. Having passed this hurdle, any given therapy has now to prove its efficacy in comparison to the "standard treatment" which is basically focused on phase III trials, be they multi or single institutional studies. Multiple and very diverse requirements during the different steps of evolution of any new therapeutic concept need different model systems. This diversity elucidates the impact of a proper choice to select the appropriate model system. The following short review wants to give a rough outline regarding the armamentarium of model systems in neurooncology.

In Vitro Studies

Most in vitro studies are done with cell cultures. The choice of the appropriate cell line is the first step towards a successful study. In general, cell lines are highly artificial systems which mainly serve to decipher a very circumscript (ideally monocausal) phenomenon on cell/biological levels. Most widespread is the use of human cell lines or cell lines from rodents (usually rat or mouse). However, rodent cell lines and human cell lines frequently require different growth conditions; moreover, they have different metabolisms and display different antibodies on the cell surface. In addition, even at the very beginning, the whole concept of the project with possible forthcoming steps have to be considered: if the use of animal models is anticipated, the choice of the proper cell line from the beginning is of critical importance. If rodent models for instance are planned, one should stick to rodent cell lines at the beginning as well. Any switch of cell line in between may significantly reduce the conclusions which can be drawn.

Culture conditions may differ for different cell lines. The impact of the micromilieu of the cell lines/cell cultures is frequently underestimated. The typical growth curve of a cell culture has a sigmoid plot. After an initial lag phase in which the cells adapt to the conditions, the proliferation follows a logarithmic curve. Afterwards, the cells level off to a plateau in which the cells have reached the final density of population prior to increasing cell loss due to nutritional limits. Experiments focussing on cell proliferation have to be done in the log phase of the maximum proliferation. Phenomena which require cell/cell contacts should be done in the very narrow time frame of the plateau phase. The time frames as well as the doubling time have to be analyzed for each cell line in every lab again. The growth conditions of a cell culture are mainly determined by growth media. The requirements for certain experiments and the need for specific culture conditions have to be adapted individually. Most cell lines and almost all primary cultures need addition of serum to the culture conditions. Most frequently fetal calf serum is used, some cells (e.g. medulloblastomas) require human serum [16]. The concentration of serum used in the culture media ranges from 2.5 to 20%! Growth factors, hormones and other inductors of the proliferation are part of this serum (albeit in different proportions and concentrations). Thus, in experiments focussing on the impact of growth factors or the regulation of growth control and cell proliferation, the culture conditions have exactly to be defined in advance. If the experiments deal with the analysis of factors created by the tumour cells themselves, serum free culture conditions are frequently needed. Cell lines and even more primary cell cultures have to be adapted hereto, which again changes the biological properties of the cells themselves. In general, cell lines and primary cell culture out of tumour biopsies have very different biological properties. Cell cultures made from primary tumour biopsies are much closer to the real in situ system [2, 5]. However, due to tumour heterogeneity (especially in case of gliomas) they are much less standardized. Therefore all cell culture modalities have to be defined and closely monitored throughout the experiments in order to get reproducible results.

Growth conditions of cell cultures are in general highly artificial compared to the situation in situ. Classical "cell culture systems" are grown as a flat monolayer on a plastic dish, the monolayer being a nearly two dimensional system. These monolayers display a thickness of one single cell. Cell/cell contacts are missing for a long period of time. Once the cells have proliferated to a certain density which allows cell/cell contacts, these are only possible in a "lateral distribution": on bottom of the cell is the culture vessel, on top of it the growth medium. Since the extracellular matrix is of utmost importance in terms of differentiation of the cells as well as cellular motility and intracellular signal transduction, coating of the cell culture dish with matrix components has been established. In addition, the growth medium can be fortified with matrix components as well [6, 11]. However, this does not mimic three dimensional cell growth [19]. Therefore three dimensional cell culture models have been introduced. Most commonly used are spheroid cultures which can be derived from suspension cultures using roller flasks, gyro flasks, or simply by seeding suspension cultures on top of a medium agar [2]. Another system is to incorporate tumour cell suspensions into semi-solid gels to allow the cells to expand three-dimensionally [18]. However, all these culture systems are artificial by themselves and require standardized procedures to allow reproducible experiments as well as the establishment of internal standards in each laboratory to define the experimental frame and to pinpoint the biological feature of interest.

Since even three dimensional culture systems do not reproduce the micromilieu of the organism, organ slice cultures have been established and have been addressed as "organotypic cell culture models". Despite very sophisticated techniques to maintain organ slice cultures, autolytic changes occur within short time. Thus it is almost impossible to get stable cell culture conditions over longer periods of experimental set up. In spite of these technical problems organ slice cultures deserve increased attention since this type of model will overcome a lot of drawbacks inherent in previous model systems. The interaction of tumour and host tissue can be studied by means of confrontational cultures. Tumour tissue and host tissue are co-incubated. This can be achieved by superimposing two monolayers of different origin, by confronting spheroids of tumour and host tissue or by adding a cell suspension/cellular spheroid on top of the organ slice [3]. Hereby the interaction of different tissues can be studied, especially the critical phenomena of cell migration and tumour cell invasion. Cellular migration (directional motility of cells) can be quantified by adding defined amounts of a certain cell population on standardized culture dish devices be they coated or uncoated. Within defined time frames (usually 48 hours or less), the distance of migration can be measured and the mean cellular velocity can be calculated [4, 13, 21]. Hereby it is possible to study the effect of different substances on cellular migration. In the three dimensional space, the same is possible in semi solid gels. Semi solid collagen gels have proven to be very reproducible. In terms of obtaining data, cells can even be monitored by time lap video microscopy. However, the consistency of the gel (often dependent on the collagen concentration) has a critical influence on the cellular motility [18]. Therefore, for different cell types the individual experimental set up has to be optimized.

Another widespread system is the "Boyden Chamber Assay". A porous filter (with a pore size usually of 8 micrometer) can be used either uncoated or, in most cases, coated with matrigel. Matrigel is an artificial mixture of different components of the basement membrane. A dual chamber system is separated by these filters with the cells under investigation being on top of the filter and a "chemoattractant" in the lower compartment. Hereby the proportion of cells penetrating the filter can be quantified under different experimental conditions [15].

The invasion of cells into solid tissue can be quantified using co-cultures of different tissues of tumour/host confrontation or by putting cells on top of the organ slice culture. The distance of penetration of the cells under investigation can be quantified during a certain standardized time frame. However, detection of the cells of interest has to be performed by labour intensive histological techniques; identification of the cells frequently requires immunohistochemical techniques [3, 4, 6, 12]. Prior to any experiment the standards of experimental set-up have to be outlined using several positive and negative controls – without extensive control systems these systems turn out to be invalid.

In Vivo Systems

The next step after successful evaluation of a hypothesis in vitro is to establish an appropriate in vivo model. There are numerous in vivo models available in neurooncology, however, the appropriate choice is a major prerequisite for the success of the study. Usually, rodent models are chosen because these animals are rather easy to handle and available in large quantities. However, many possible pitfalls have to be kept in mind:

Anatomy and physiology in rodents are quite different from larger animals or humans. Rats and mice have by far less white matter in proportion to grey matter areas. Since gliomas extend into the white matter following rather distinct pathways of invasion, the lack of an appropriate amount of white matter in rodents can be a major drawback for a study. Moreover, a tentorial area with a sharp distinction between supra- and infratentorial compartments is missing in rats. Thus the effects of mass shift are of minor relevance for a long time, allowing tumours to grow extensively. Many rodent tumour cell lines which are available for intracranial implantation grow with a rather sharp delineation, which is in contrast to the diffuse invasive character of human gliomas in situ. Recently a model has been reported implanting human glioma biopsy spheroids or biopsy fragments into the brain of nude rats. This results in rather diffuse tumour invasion mimicking the clinical situation in humans [7]. However, nude rats are very tricky to handle, thus available only in a few places. Moreover nude rats are highly artificial systems due to the lack of a regular immune system. Many cell-lines have been used as transplants into rat brains. The most common line, C6 seems to provoke an immunoresponse in the animals. This observation severely limits the usefulness of this cell-line for longterm studies, especially concerning longterm survival in treated versus untreated groups [19]. Another cell-line marker which was used for a long time to track cells in organ systems was lacZ. LacZ induces a severe immune response which strongly interacts with most experimental set-ups. Therefore the use of lacZ seems to have a very limited place in this respect [8, 22]. The choice of appropriate tumour cell-lines to be implanted into an animal's brain cannot be overemphasized. With regard to identification of either secreted proteins or cellbound proteins, one has to clarify in advance whether or not the antibodies used to identify the protein of interest are working specifically enough in the species of choice. Numerous monoclonal antibodies have a very limited spectrum whereas cross-specificity has to be tracked for each individual antibody in the anticipated model system prior to the start of large series of animal experiments.

In order to mimic metastatic disease, cell suspensions have been injected into the general circulation. Using appropriate amounts of well characterized cell suspension, this is a rather standardized system for a metastatic disease.

In order to monitor certain biological effects in vivo, sophisticated techniques have been developed recently. Modern MRI technology allows to visualize tumour growth even in small rodent systems with high precision. This can be accomplished with standard MRI systems and common software equipment [9]. Intravital microscopy offers additional insight into tumour physiology especially with respect to microcirculation. However, skinfold chamber models are only a first step since orthotopic location of the tumour has to be favoured (e.g. cranial window technique) [20].

Large animal models usually use pigs or dogs. Certain techniques of application (e.g. biodegradable wafers, cells encapsulated in alginates) require large animal systems. Meanwhile a canine glioma model has been successfully established, albeit requiring a very complex experimental set-up [1, 17].

Altogether, since the choice of the appropriate animal model is of critical importance with regard to the meaningful result of the project, the value of intense exchange of information between groups working with animal models cannot be overemphasized.

Clinical Projects

The guidelines for the conception of clinical study protocols are not within the scope of this short overview. However, some remarks concerning clinical projects seem to be appropriate since the first steps into the clinic coming from the animal lab to bedside still have a "model character". Once in the situation to prove the anticipated clinical effect of a system dissected so far in vitro and in animal studies, one has to critically ask what kind of patients again serve as "optimized models". Usually recurrent tumours with no other therapeutical options are chosen. This attitude, mostly driven by ethical considerations, might be of major disadvantage for the hypothesis to be tested. Patients/tumours having been pretreated by numerous therapies are often refractory to any other further therapy, whatever it may be. Moreover, these heavily pretreated patients are "biological systems" underlying multifactorial treatment influences and it may be at least not easy, in most cases even impossible, to dissect any effect one would like to identify in this study cohort. With regard to this limit, any therapeutic study which cannot selectively dissect its target, might be inethical per se even in this group of patients having no other choice of treatment. Whereas the usual dose escalation/toxicity studies might be feasible in this cohort of patients, the evaluation of any effectiveness might be much more meaningful (and in this respect much more ethical) in a group of patients being treated upfront unless therapeutic concepts with proven efficacy are not withheld. In any case, a very thoughtful systematic analysis is necessary to identify the effect of a given novel therapy and to define what amount of effect is of real benefit for a given patient (although being a statistically significant result, a prolongation of survival of about 3–4 weeks in a disease lasting for 2 years is questionable to be a *real satisfactory* effect). In this respect, large multicenter trials have to be planned extremely carefully in order to get a satisfactory balance of effort and impact. Only the honest and critical analysis of very well designed studies discovers results which give credit to all the work and effort that have previously been put in all modelling systems within a project.

References

1. Berens ME, Giese A, Shapiro JR, Coons SW (1999) Allogeneic astrocytoma in immune competent dogs. Neoplasia 1: 107–112
2. Bjerkvig R, Tonnesen A, Laerum OD, Backlund EO (1990) Multicellular tumor spheroids from human gliomas maintained in organ culture. J Neurosurg 72: 463–475
3. Engebraaten O, Bjerkvig R, Lund-Johansen M, Wester K, Pedersen PH, Mork S, Backlund EO, Laerum OD (1990) Interaction between human brain tumour biopsies and fetal rat brain tissue in vitro. Acta Neuropathol 81: 130–140
4. Bjerkvig R (ed) (1992) Spheroid culture in cancer research. CRC Press, Boca Raton Ann Arbor London
5. Bouterfa H, Picht T, Kess D, Herbold C, Noll E, Black P Mcl, Roosen K, Tonn JC (2000) Retinoids inhibit human glioma cell proliferation and migration in primary cell cultures but not in established cell lines. Neurosurg 46: 419–430
6. Engebraaten O, Bjerkvig R, Pedersen PH, Laerum OD (1993) Effects of EGF, bFGF, NGF and PDGF (bb) on cell proliferative, migratory and invasive capacities of human brain tumour biopsies in vitro. Int J Cancer 53: 209–214
7. Engebraaten O, Hjortland GO, Hirschberg H, Fodstad O (1999) Growth of precultured human glioma specimens in nude rat brain. J Neurosurg 90: 125–132
8. Goldbrunner RH, Bouterfa H, Vince GH, Bernstein JJ, Roosen K, Tonn JC (1997) Transfection and dye premarking of human and rat glioma cell lines affects adhesion, migration and proliferation. Anticancer Res 17: 4467–4471
9. Goldbrunner RH, Bendszus M, Sasaki M, Kraemer T, Plate KH, Roosen K, Tonn JC (2000) Vascular endothelial growth factor-driven glioma growth and vascularization in an orthotopic rat model monitored by magnetic resonance imaging. Neurosurgery 47: 921–930
10. Goldbrunner RH, Wagner S, Roosen K, Tonn JC (2000) Models for assessment of angiogenesis in gliomas. J Neuro Oncol 50: 53–62
11. Mahesparan R, Tysnes BB, Edvardsen K, Haugeland HK, Cabrera IG, Lund-Johansen M, Engebraaten O, Bjerkvig R (1997) Role of high molecular weight extracellular matrix proteins in glioma cell migration. Neuropathol Appl Neurobiol 23: 102–112
12. Nygaard SJ, Haugland HK, Laerum OD, Lund-Johansen M, Bjerkvig R, Tysnes OB (1998) Dynamic determination of human glioma invasion in vitro. J Neurosurg 89: 441–447
13. Pilkington GJ, Bjerkvig R, De Ridder L, Kaaijk P (1997) In vitro and in vivo models for the study of brain tumour invasion. Anticancer Res 17: 4107–4109
14. Parsa AT, Chakrabarti I, Hurley PT, Chi JH, Hall JS, Kaiser MG, Bruce JN (2000) Limitations of the C6/Wistar rat intracerebral glioma mode implications for evaluating immunotherapy. Neurosurgery 47: 993–1000
15. Paulus W, Tonn JC (1994) Basement membrane invasion of glioma cells mediated by integrin receptors. J Neurosurg 80: 515–519

16. Pietsch T, Scharmann T, Fonatsch C, Schmidt D, Ockler R, Freihoff D, Albrecht S, Wiestler OD, Zeltzer P, Riehm H (1994) Characterization of five new cell lines derived from human primitive neuroectodermal tumors of the central nervcous system. Cancer Res 54: 3278–3287
17. Rainov NG, Koch S, Sena-Esteves M, Berens ME (2000) Characterization of a canine glioma cell line as related to established experimental brain tumor models. J Neuropathol Exp Neurol 59: 607–613
18. Tamaki M, McDonald W, Amberger VR, Moore E, Del Maestro RF (1997) Implantation of C6 astrocytoma spheroid into collagen type I gels: invasive, proliferative, and enzymatic characterizations. J Neurosurg 87: 602–609
19. Tonn JC, Ott MM, Bouterfa H, Kerkau S, Kapp M, Müller-Hermelink HK, Roosen K (1997) Inverse correlation of cell proliferation and expression of progesterone receptors in tumor spheroids and monolayer cultures of human meningiomas. Neurosurgery 41: 1152–1159
20. Vajkoczy P, Menger MD (2000) Vascular microenvironment in gliomas. J Neurooncol 50: 99–108
21. Vince GH, Bouterfa H, Goldbrunner R, Roosen K, Tonn JC (1997) Fast blue, a fluorescent tracer in glioma cell culture, affects cell proliferation and motility. Neurosci Lett 233: 148–150
22. Visted T, Thorsen J, Thorsen F, Read TA, Ulvestad E, Engebraaten O, Sorensen D, Yla-Herttuala S, Tyynela K, Rucklidge G, Edvardsen K, Bjerkvig R, Lund-Johansen M (2000) lacZ-neoR transfected glioma cells in syngeneic rats: growth pattern and characterization of the host immune response against cells transplanted inside and outside the CNS. Int J Cancer 85: 228–235

Correspondence: J. C. Tonn, Neurochirurgische Klinik, Klinikum Großhadern, Ludwigs-Maximilians Universität, Marchioninistr. 15, D – 81377 Munich, Germany.

Acta Neurochir (2002) [Suppl] 83: 85–91

Neuroendocrine Research Models

N. Akalan

Department of Neurosurgery, Hacettepe University School of Medicine and Institute of Neurological Sciences, Hacettepe University, Ankara, Turkey

Summary

The neuroendocrine system has always been a main research area for neuroscience. The distinctiveness of this system is due to its diverse anatomical and physiological properties, which modulates the homeostasis and reproductivity of the organism. For decades, generated research models have served to elucidate the complex properties of this system and provided invaluable data on the physiology and mechanism. Currently, the basic research has concentrated on the mechanisms of neuropeptide secretion, carriage and transportation with neurotransmitters and mechanisms operative during various physiological or pathological states. Comparable to the basic research, neuroendocrine models also help to investigate certain clinical aspects of neurological and neurosurgical diseases. Pituitary adenomas have been induced by several methods in rats to elucidate adenoma morphology, characteristics and treatment options. More recently, experimental studies are concentrated on the effectiveness of therapeutic gene transfer in neuroendocrine disorders and tumors. Moreover, animal models are of great importance for both to investigate the role of hypothalamus in aging and the potential of correcting age-related neurodegenerative processes in this setting.

Keywords: Neuroendocrine; experimental; animal models.

Introduction

The pituitary gland, stalk and certain areas of the hypothalamus constitute the neuroendocrine system, one of the major efferent systems connecting the central nervous system to the body. The neuroendocrine system utilizes the pituitary gland to regulate specific visceral organs to regulate growth and function of specific target organs. By this way, central nervous system controls and regulates the growth and development of the organism, maintains its internal milieu, regulates its metabolism, and ensures its reproduction. The neuroendocrine system as a whole is responsible for homeostasis and reproductivity and recently related to degenerative processes responsible for aging. Besides its diverse anatomy and physiology, the neuroendocrine system has been investigated intensively for its role within the central system and related pathologies.

Anatomy and Physiology, an Overview

Pituitary is the major neuroendocrine modulator that regulates many peripheral glands and tissues through the secretion of anterior pituitary hormones. It is composed of glandular and neural tissue called adeno- and neurohypophysis.

The adenohypophysis is believed to arise from the primitive foregut, the stomodeum. A diverticulum of stomodeum is thought to migrate cranially, disconnect from the foregut, and join to a neural diverticulum emerging from the diencephalon-the neurohypophysis. With the establishment of vascular connections between the neurohypophysis and the displaced foregut tissue, differentiation of the stomodeal remnant into the adenohypophysis occurs.

The adenohypophysis cells secrete eight known hormones; growth hormone, prolactin, follicle-stimulating hormone, luteinizing hormone, thyroid-stimulating hormone, adrenocorticotropic hormone, melanocyte-stimulating hormone and beta-endorphin. The adenohypophysis is divided into three regions, the pars tuberalis, pars intermedia, and pars distalis. The pars tuberalis rests at the surface of the median eminence and the upper infundibular stem, the rostral regions of the neurohypophysis. It is made up of epithelial cells, fenestrated capillaries, and stromal cells. No nerve terminals are present in the pars tuberalis. The epithelial cells have been identified as thyrotropes and gonadotropes. In many species, a pars intermedia is present.

It is made up of epithelial cells, with only a few capillaries and stromal cells. Dopaminergic nerves are found in the pars intermedia and terminate near glandular cells. Immunohistochemical and physiological studies have demonstrated that its epithelial cells contain a-MSH and beta-endorphin. The pars intermedia is present in the fetal human and in the pregnant adult female but is absent in adult human males and nonpregnant females.

The pars distalis forms the bulk of the adenohypophysis. It is composed of epithelial cells arranged in a glandular pattern, stromal cells, and fenestrated capillaries without any axon terminals. Several studies have demonstrated that lactotropes and somatotropes lie predominantly in the lateral wings of the pars distalis, whereas thyrotropes and gonadotropes lie in the medial third, so called the "mucoid wedge" because the secreted hormones contain glycoproteins. As this median zone is continuous with the pars tuberalis and contains similar functional cell populations, it is also termed the zona tuberalis. Corticotropes lie anteriorly in the mucoid wedge and over the surface of the lateral wings. Melanotropes lie posteriorly near the neural lobe, with a small number scattered throughout the pars distalis.

Pituitary hormones are synthesized on rough endoplasmic reticulum. The sequence of amino acids making up the hormones is encoded in nuclear DNA. Through the mediation of messenger RNA (mRNA), this information is transferred to the cytoplasm. Protein synthesis occurs in the cytoplasm with the interaction of mRNA with ribosomes, with the mRNA dictating the sequence of amino acids. The produced protein is transported to the Golgi complex and packaged in large vesicles.

The neurohypophysis is a diverticulum of brain, which appears in the human early in fetal life. The mature neurohypophysis is made up of axon terminals, specialized glial cells, and blood vessels. The neurohypophysis contains no neuronal cell bodies, only axons and axon terminals. Axons terminate in the perivascular space of fenestrated capillaries, not on neurons or their processes. The neurohypophysis lacks a blood-brain barrier. It regulates the function of the adenohypophysis. The neurohypophysis is subdivided into three regions on the basis of morphological specializations, the median eminence, the infundibular stem, and the neural lobe. The median eminence, with the paired lateral eminences, constitutes the tuber cinereum, lying caudal to the optic chiasm and rostral to the mamillary bodies. As the median eminence forms the funnel-shaped floor of the third ventricle, it is also called the infundibulum. The infundibular stem is the neural portion of the pituitary stalk. The neural lobe (infundibular process) is the caudal region of the neurohypophysis. The classification of the neurohypophysis into the infundibulum (median eminence), infundibular stem, and infundibular process (neural lobe) emphasizes the observation that the neural portion of the pituitary gland is a diverticulum of brain but is distinct from the hypothalamus. The infundibulum is separated into an ependymal layer, an internal zone, and an external zone. Its ependymal layer is made up of specialized epithelial cells. They are united by tight junctions which inhibit the passive exchange of materials between the third ventricle and the interstitial fluid of the infundibulum. These cells lack cilia at their ventricular surface. In some regions their apical surface is characterized by the presence of large blebs as well as numerous smaller microvilli. Some of these ependymal cells ("tanycytes") are stretched, with their apical surface facing ventricular fluid and their basilar processes terminating in the perivascular space of fenestrated capillaries on the median eminence surface.

The median eminence internal zone is made up of axons of the supraopticohypophyseal tract, originating from hypothalamic supraoptic and paraventricular nuclei and pass through the median eminence to terminate in the neural lobe. In addition, noradrenergic fibers and terminals have been demonstrated in the internal zone. These fibers are believed to originate outside the hypothalamus in the brain stem and to represent the terminus of the ascending reticuloinfundibular tract. The median eminence external zone is made up of glial cells, axons, and axon terminals. Light microscopic studies demonstrate that the dopaminergic tuberoinfundibular tract originating in the hypothalamic tuberal nuclei terminates in this region.

The infundibular stem lies between the infundibulum and the infundibular process. It is characterized by the presence of axons of the supraopticohypophyseal tract. The dopaminergic tuberohypophyseal tract is also present in this region.

The infundibular process (neural lobe) is the terminus of the supraopticohypophyseal tract. The axons of the supraopticohypophyseal tract terminate in the perivascular space of neural lobe capillaries. Near the adjacent adenohypophysis, terminals of the dopaminergic tuberohypophyseal tract may also be found.

The secretory activity of the gland is controlled by signals derived from the hypothalamus in the form of releasing and inhibiting hormones and neurotransmitters. The hypothalamic nuclei differ from other neuronal circuits in terms of communication with the end-organ, the pituitary. These so-called neurosecretory cells receive information from the body like all other neurons but generate secretion instead of impulses, use specified vascular system, altogether making them unique within the central nervous system. The secretory epithelial cells of the pituitary are apart from the central neural elements which control their function. The final common pathway to these endocrine cells, unlike the other efferents of the central nervous system, is not neural but vascular. The information is carried by neurohormones released in a pulsatile pattern from neuro-hypophyseal axon terminals by the process of neurosecretion into restricted vascular channels passing to the adenohypophysis. Information in the form of neurosecretions from axon terminals is carried from the neurohypophysis to the adenohypophysis by vascular routes. The adenohypophysis does not receive a direct arterial supply. Blood entering the adenohypophysis first passes through the neurohypophysis. The venous drainage of the adenohypophysis is from the pars distalis to the adjacent cavernous sinus. The primary drainage routes are through the same posterior hypophyseal veins which drain the neurohypophysis.

Hypophysiotropic hormones of the hypothalamus are synthesized in the cell soma on rough endoplasmic reticulum by the same series of steps involved in pituitary peptide synthesis within adenohypophyseal cells. The sequence of amino acids is determined by nuclear DNA, and the information is carried to the cytoplasm by messenger RNA (mRNA). The prescribed sequence is formed by the interaction of mRNA with ribosomes. The translated protein hormones are transported to the Golgi apparatus and packaged in large, granular vesicles. Vesicles containing the hormones are transported by axoplasmic flow from the cell soma to the axon terminal lying in the perivascular space of a fenestrated capillary in the neurohypophysis. Recent evidence suggests that the hormone translated from mRNA is frequently a prohormone considerably larger than the final biologically active product. Post-translation modification of the prohormone by cleavage of peptide bonds at selective sites occurs during axoplasmic transport of hormones within vesicles.

Two peptidergic neurosecretory systems terminate in the neurohypophysis: the first is the magnocellular system with cell bodies in the hypothalamic supraoptic and paraventricular system. This system projects to the infundibular process (neural lobe) through the supraopticohypophyseal tract. Cells in this system contain oxytocin and vasopressin and their associated neurophysins. Oxytocin and vasopressin are found in both the supraoptic and paraventricular nuclei but are found in different cells within those nuclei. As noted above, the supraopticohypophyseal tract forms the internal zone of the median eminence. The cells and axons are typically large, characterized by the presence of large dense core vesicles which contain either oxytocin and neurophysin I or vasopressin and neurophysin II. The second neurosecretory system is the parvocellular system. The cells in this system originate in several hypothalamic cell groups and terminate in the median eminence in the perivascular space of fenestrated capillaries. Its terminals are mostly found in the median eminence external zone. The neurons are typically small, and their axon terminals contain small synaptic and large granular vesicles. By immunohistochemical techniques, GnRH, TRH, SOM, and CRH have been localized within cells of the parvocellular peptidergic system.

Evolution of Neuroendocrine Research

Initial studies on neuroendocrine system have been more or less concentrated on elucidating basic relationship and physiology. Studies beginning from the 1900s provided invaluable information on basic concepts of this system.

Hypophyseal hormones and their function was, to a great extent, due to the pioneering studies of Cushing and associates who defined the pituitary gland as "conductor of the endocrine orchestra". These workers were the first to demonstrate the pituitary gland, while not essential for life, normally exerts an important influence on the metabolic processes of the body. Subsequent studies revealed the importance of pituitary function in the development of sexual organs and behavior. Once the framework of this system has been clarified, the second step was to establish the link to the central nervous system. Transplantation of the pituitary gland has been widely performed especially to investigate its physiological properties and demonstrate its dependence on hypothalamic influences [1, 6, 7]. Experiments starting at the turn of the century have shown that pituitary tissue, when transplanted to remote sites other than the anatomic position, partly perform its secretory function. Initial attempts in which the anterior chamber of the eye and the renal capsule were chosen for transplantation sites have proved that the pituitary graft should be anatomically bound to the hypothalamus [11, 15]. The clinical relevance of these studies were that the diminished pituitary function in humans for various reasons can only partly be restored by pituitary tissue grafted to kidney capsule or subcutaneous tissue. One of the remarkable studies on the neuroendocrine system of the rat belongs to G. W. Harris and D. Jacobson [12] in 1952, where they had demonstrated that the pituitary tissue can function much better in its original site than remote parts. Their results allowed them to state that the pituitary function is not only controlled through the vascular route. Soon after, M. Nikitovitch-Winer and J. W. Everett [17] in 1958, showed that proper pituitary tissue function depends on the anatomic contact to the hypothalamic area. Their experimental design was to remove the transplanted pituitary tissue from the renal capsule in the hypophysectomised rat and re-transplanted it to the anterior hypothalamic area. Based upon the histological changes of the end-organs for the neurohypophyseal system, they had concluded that the grafts functioned much better when they are in close proximity to the anterior hypothalamus. It was not before 1977 that Schally and Guillermin, in their Nobel Prize winning study, showed that certain hypothalamic releasing and inhibiting hormones and factors are responsible for controlling the pituitary tissue for humoral secretion. Availability of the radioimmunassay methods and advancement of electron microscopy methodology made it possible to demonstrate the ultrastructural and biochemical evidence of the restored pituitary function after transplantation [1]. Subsequent work by Schally and co-workers and by Guillemin and Peterson and by others led to the detection of a number of hypothalamic factors controlling the release of pituitary hormones. Thyrotropin-releasing hormone (TRH) was identified and biochemically analyzed in 1969; gonadotropin-releasing hormone (LH-RH) in 1971, growth hormone release-inhibiting hormone (GH-RIH) in 1973, corticotropin-releasing hormone (CRH) in 1981–82, and growth hormone-releasing factor (GH-RF) identified in 1965 and characterized in 1980–83.

The endocrine function of the hypothalamus has been the subject of several recent reviews. As the knowledge on the anatomy and the physiology of the system have improved, the research models aimed at the hypophysotrophic area of the hypothalamus; after the 70s, for the possibility of restoring reduced function at deficiency states like diabetes insipidus or hypogonadism [5]. The experimental studies of this period represent the second phase of neuroendocrine research, therapeutic approach to neuroendocrine diseases. Besides the fundamental data obtained by experimental central nervous system transplantation, the idea of restoring hypothalamo-hypophyseal deficiencies in human beings by pituitary tissue replacement has formed the basis for a neurosurgical approach to pituitary transplantation. Wilberger [20], not only gave excellent documentation of the central nervous tissue transplantation studies but also proposed transplants of the pituitary-hypothalamus in other pathological conditions. Parallel to the on-going studies on experimental central nervous system transplantation, data obtained in those studies revealed that hypothalamic tissue transplants could survive and were able to restore diminished hypothalamic function [1, 4, 5, 13].

Besides hormonal restoration, transplantation studies on the neuroendocrine system contributed to another important issue of neural transplantation, to explore the differences between the regenerative capacity of central and peripheral nervous tissue. The discrepancy between the regenerative capacity of the peripheral and central nervous system has been a major area of interest. Inability of the central nervous system to regenerate has been attributed to the lack of certain trophic factors that have been identified in the peripheral but not in the central nervous tissue [16]. The presence of certain factors which are supposed to be operative during embryonic period but become inactive later in life in the central nervous system is another topic of interest. Hypothalamo-hypophyseal system, with its extraordinary properties provide a unique milieu for experimental regeneration studies. Although the hypothalamus and the posterior lobe of the pituitary represent the central nervous system in terms of regenerative capacity, vascular structure in terms of blood-brain barrier (BBB), anterior lobe and median eminence along with the portal system show the properties of the peripheral nerve tissue. This discrepancy of basic properties has provided the unique advantage of carrying the regenerative capacity of the peripheral tissue to damaged sites in hypothalamus, hoping to achieve better regeneration through transplantation. Studies enabled to speculate that end-organ (pituitary) promotes regeneration by either providing a leaky BBB or by generating growth factors, when transplanted to lesioned areas of the anterior hypothalamus [2, 14].

Fundamentals of Neuroendocrine Research

As seen from the historical evolution summarized above, neuroendocrine research models served several purposes: First, anatomy and physiology based, aiming to elucidate the extraordinary properties of the neuroendocrine system described above. Second, to disclose pathophysiology of diseases, due to developmental disorders, tumors or degenerative processes involving the hypothalamo-hypophyseal system. Finally, research on the neuroendocrine system aims to provide scientific basis to restore the deficiencies or reverse the pathological changes of disease of this system.

Constructing an Experimental Design

From the neurosurgical standpoint, construction of an experimental design in the neuroendocrine system follows the same route as any area of research. The first step is to create a hypothesis and an experimental model that would be comparable to human, in terms of anatomy, physiology or pathology. As observed in the literature, rat has been one of the most preferred mammal for neuroendocrine research models. It is relatively cheap, easy to obtain and has a well documented anatomy and physiology.

Especially when investigating a deficiency state or pathological condition, the second step should be creating a model for the desired pathological condition. For searching anatomical and physiological characteristics or pathophysiology of the system, ablative procedures can establish the necessary condition in the laboratory animal. In the neuroendocrine system, surgical hypophysectomies are the most common ablative procedures performed and various methods have been described for the rat [8].

Animal models demonstrating specific deficiency states of the hypothalamic area is also available. It is possible to reach certain targets in the hypothalamus with the aid of stereotactic methods to obtain lesions at the desired neuronal populations. By this way, releasing or inhibiting hormones to the pituitary may be arrested. Ablation technique for the hypothalamic area requires a stereotactic system specific to the animal along with stereotactic atlas for calculating the appropriate targets [18]. Ablative procedures can be performed either mechanically or by delivering neuron specific toxins to the desired area. Both hypophysectomies and lesion generation at the hypothalamus require considerable effort and workup at the beginning.

Tissue transplantation techniques also have implications in creating models for neuroendocrine research. Transplantation techniques have the advantage of providing an in-vivo environment for research on both anatomy and physiology of the system, regenerative properties or pathological conditions such as tumors. Stereotactic methods can efficiently be used for the transplantation of fetal pituitary or hypothalamic tissue, certain pituitary tumor lines to the desired localization for various purposes.

Genetically deficient strains of laboratory animals also comprise novel models. Certain strains of rats with congenital deficiency of certain hypothalamic hormones serve as excellent models for hypothalamic research. Brattleboro rats with congenital antidiuretic hormone (ADH) deficiency and hypogonadal mice with gonadotropic hormone releasing hormone (GHRH) deficiency have been used in numerous studies.

Whatever the chosen method for creating a research model may be, most important issue is that it should be reproducible. Only reproducible models can provide valuable data, allowing others to test and improve the findings.

Particularly for those aiming at clinical applications for neuroendocrine disorders, third tier of a research project is to establish a method of reversal of the pathological condition created in the model. This can be done by injection of certain elements, transplantation of tissue or cell populations to substitute the missing ones in the model, delivering drugs by various methods and finally by gene transfer.

The final part of the experiment is the assessment of the results. The results are based upon various data obtained according to the experimental design. The data may be based on morphological changes; biochemical measurements or clinical criteria depending on the aim and method of the research. It is crucial to determine on what and how to look for. The availability of a control group, obtaining enough and comparable data with proper statistical evaluation is not only the basis of neuroendocrine but any type of research, as well.

Current Areas of Interest

All these studies, although providing invaluable basic data on the physiology and mechanisms of the neuroendocrine system in health and disease, clinical implication was not as practical, because all pituitary hormones with their releasing and inhibiting hormones in the hypothalamus are already available either in natural or synthetic forms to ameliorate the dysfunction. Instead, the research on the neuroendocrine system has once more focused on physiology of the neuropeptide secretion, carriage and transportation with neurotransmitters and mechanisms operating during various physiological or pathological states [9].

Besides the deficiency states of the neuroendocrine system, experimental models have also been generated to investigate the morphology, characteristics and treatment options of pituitary tumors [19]. These studies have been conducted by inducing specific types of pituitary tumors in laboratory animals, such as estrogen induced adenomas in rats that served to explore the effectiveness of different treatment modalities. Certain pituitary tumor bearing rat models have also been available. Wistar/Furth rats resembling human prolactinomas, for example, have been utilized for the same purposes. These models have served to investigate several aspects of tumorigenesis and treatment options with limited success.

Most recently, experimental studies are concentrated on assessing viral gene therapy in neuroendocrine models. Development of powerful new viral-based gene transfer systems has generated a great deal of research interest in the field of therapeutic gene transfer in neuroendocrine disorders and tumors [10]. Current research efforts include treatment of experimental pituitary tumors by adenoviral vector-mediated transfer of the suicide gene for the HSV-1 thymidine kinase which converts the prodrug ganciclovir into a toxic metabolite. At the hypothalamic level, an adenovirus harboring the cDNA for arginine vasopressin has been used in Brattleboro rats to correct diabetes insipidus for several weeks [3].

Aging and hypothalamus have also been a certain aspect of research, and neuroendocrine models have served to delineate the role of hypothalamus in aging. It is well established that destruction of the ventromedial nucleus (VMN) in the hypothalamus produces signs and symptoms that mimic age related changes. Rat models with VMN lesions will enable to investigate the theories of aging especially related to hypothalamic nuclei, the role of neurotransmitters including neuropeptide Y, beta-endorphin and corticotrophin releasing hormone in the process of aging. Using these models, the potential of gene therapy to correct age-associated neurodegenerative processes can be explored on neuroendocrine cellular level.

Research models have served as indispensable tools to improve our knowledge on basic anatomy and physiology in the neuroendocrine system. With advancing technology, they will continue to provide information for basic science at the molecular level. Furthermore, these models will continue to be developed to better understand the cascade of events in pathological conditions and search for probable solutions.

References

1. Akalan N, Pamir N, Benli K, Erbengi A, Erbengi T (1988) Fetal pituitary transplants to the anterior hypothalamic area of the hypophysectomised rats: an ultrastructural and histochemical study. Surg Neurol 30: 342–349
2. Akalan N (1996) Effects of neural lobe implants on the regeneration of neurosecretory axon regeneration in hypothalamo-neurohypophyseal system lesions. Doctoral thesis, Institute of Neurological Sciences, Hacettepe University
3. Bolognani F, Goya RG (2001) Gene therapy in the neuroendocrine system: its implementation in experimental models using viral vectors. Neuroendocrinology 73(2): 75–83
4. Carithers J, Dellmann HD (1992) Regeneration of neural lobe-like neurovascular contact regions in explanted neural lobes placed in the hypothalamo-neurohypophysial tract in the lateral retrochiasmatic area. Brain Res 2; 592(1–2): 63–73
5. Charlton KM (1992) Hypothalamic transplantation. Ciba Found Symp 168: 268–275
6. Diamond MC (1955) The effect of hypophysectomy, removal of the posterior pituitary, adrenalectomy and dehydration on the antidiuretic activity of the rat hypothalamus. Endocrinology 58: 461–470
7. Everett JW (1956) Functional corpora lutea maintained for months by autografts of rat hypophysis. Endocrinology 58: 786–796
8. Falconi G, Rossi GL (1964) Transauricular hypophysectomy in rats and mice. Endocrinology 74: 301–303
9. Flanagan-Cato LM (2000) Estrogen-induced remodeling of hypothalamic neural circuitry. Front Neuroendocrinol 21(4): 309–329
10. Geddes BJ, Harding TC, Lightman SL, Uney JB (1999) Assessing viral gene therapy in neuroendocrine models. Front Neuroendocrinol 20(4): 296–316
11. Goldberg RC, Knobil B (1957) Structure and function of intraocular hypophyseal grafts in the hypophysectomised male rat. Endocrinology 74: 301–303
12. Harris GW, Jacobson D (1952) Functional grafts of the anterior pituitary gland. Proc R Soc Biol 139: 263–267
13. Jain S, Mathur R, Sharma R, Nayar U (1999) Neural tissue transplant in the lateral hypothalamic lesioned rats functional recovery pattern. Neurobiology (Bp) 7(4): 421–430
14. Jezova D, Johansson BB, Oprsalova Z, Vigas M (1989) Changes

in blood-brain barrier function modify the neuroendocrine response to circulating substances. Neuroendocrinology 49(4): 428–433
15. Knigge KM (1962) Gonadotrophic activity of neonatal pituitary glands implanted in the rat brain. J Physiol 202: 387–391
16. Kromer LF, Cornbrooks CJ (1985) Transplants of Schwann cultures promote axonal regeneration in the mammalian brain. Proc Nath Acad Sci 82: 6330–6334
17. Nikitovitch-Winer M, Everett JW (1958) Functional restitution of pituitary grafts re-transplanted from kidney to median eminence. Endocrinology 63: 916–931
18. Pellegrino LJ, Pellegrino AS, Cushman AJ (1979) A stereotaxic atlas of the rat brain, 2nd edn. Plenum Press, New York
19. Trouillas J, Girod C (1988) Animal models of human pituitary tumors. Pathol Res Pract 183(5): 638–641
20. Wilberger JE (1983) Transplantation of central nervous tissue. Neurosurgery 13: 90–94, discussion 275–286

Correspondence: N. Akalan, Department of Neurosurgery, Hacettepe University School of Medicine and Institute of Neurological Sciences, Hacettepe University, Ankara, Turkey.

Acta Neurochir (2002) [Suppl] 83: 93–99

Restorative Neurosurgery

S. Palaoğlu

Neurosurgery Department, Hacettepe University, School of Medicine, Ankara, Turkey

Summary

Restorative neurosurgery currently is the frontier of neuroscientists for the restoration of lost neuronal function especially in neurodegenerative diseases and ischemic and traumatic central nervous system (CNS) disorders. The striking developments in molecular neurobiology and bio-technology are progressively offering new opportunities for a better quality of life to patients suffering from loss of neuronal function. Besides all new and challenging medical therapeutic interventions, great emphasis is also given to transplantation for neuronal restoration as well.

Keywords: Restorative neurosurgery; transplantation; neurodegenerative disease; ischemic neuronal injury; traumatic neuronal injury.

Introduction

Historically, neurosurgical interventions such as decompression for tumour, trauma and abscess; clipping of aneurysms; repairing congenital malformations; shunting hydrocephalus are all known to have a great impact on the improvement of the environment of the nervous system to promote maximal spontaneous recovery of function. Based on the combination of new technologies and scientific developments in neuroscience, restorative neurosurgery currently is advancing the frontiers of our neuroscientists providing the potential to restore lost function.

The population of patients whom we treat at present is a small portion of those who suffer from disabling neurological illnesses. There are still problems with irreversible neurological deficits, cell death and decreased number of neurons, where even the great development in neurosurgery towards microneurosurgery hasn't yet been able to reach an absolute curative solution. On the other hand, it is the goal of neurosurgeons that the enormous amount of neurobiological and bio-technological research undertaken will broaden the scope of therapy for such diseases, and patients suffering from Parkinson's disease, Alzheimer disease, Huntington's disease, multiple sclerosis, ischemic brain injury and trauma will have the chance for a somewhat high quality survival. Thus, restorative neurosurgery is expected to bring forth the opportunities for restoration of neuronal function in acquired, degenerative or idiopathic neurological diseases.

Neural Plasticity, Regeneration, Neuroprotection

Neural plasticity is defined as the capability of the nervous system to adapt to a changing internal or external environment, to previous experience, or to trauma. It is known to be the essential and central feature of adaptation. Nervous system plasticity is of great importance in relation to a significant number of health-related problems, such as peripheral nerve, spinal cord and brain injury, developmental disorders, learning disabilities and dementia. It is clear that $Ca2+$ homeostasis and $Ca2+$ mediated cellular responses are of significance to neuronal plasticity during development, adulthood and ageing.

Damage to the central nervous system causes damage to neurons. This damage causes death of neurons since axons are not able to regenerate in the mammalian brain. Although intervertebrates and lower vertebrates are quite able to regenerate their axons in their central nervous system, this ability has been gradually lost during evolution and in mammals is almost completely absent [1].

Recently, much of the interest in the field of axon regeneration has focused on the role environment in encouraging or inhibiting axon regeneration. There appear to be three main reasons why axons don't regrow in the CNS: the inhibitory factors on the surface of oligodendrocytes, the relative impenetrability of astrocytes, and the declining vigour of axon regeneration with age. It has been suggested that CNS tissue is inhibitory and that essentially all axons have the ability to regenerate. Thus, if peripheral nervous tissue is implanted into the brain of animals, regeneration into peripheral nervous tissue occurs. In the peripheral nervous system the critical cell type is the Schwann cell for regeneration capacity and if these cells are removed from or excluded from the region where axons are trying to regenerate, there is no growth [19].

However, the inhibitory influences of both astrocytes and oligodendrocytes are removed when peripheral nerve graft is inserted into the CNS. Such findings are explanatory for the regeneration of the axons of mature neurons only within the Schwann cell environment, and removal of the two-glial effects may in itself be important factors for regeneration capability [9].

Moreover, for promotion of regeneration in the CNS, it is suggested that an appropriate extracellular substrate must be provided; the inhibitory molecules produced by a subtype of oligodendrocyte-like cells must be neutralised and appropriate neurotrophic factors must be provided. It is known that several endogenous molecules affect neuronal plasticity. Thus, any kind of intervention parallel or against these endogenous molecules is all expected to affect neuronal plasticity towards a restorative status. Calcium is known to have the most important role in neuronal physiology and thus Ca^{2+} channel antagonists, especially long lasting (L-Type) channel antagonist, dihydropyridine compounds, are still experimentally and clinically studied for their possible neuroprotective effects. Free Oxygen radical-mediated lipid peroxidation is one of the major mechanisms of secondary damage in traumatic brain injury. It has been shown that nimodipine produces neuroprotective effects not only because of its calcium channel blocking activity but also because of its inhibitory effect on lipid peroxidation, also when given in the acute phase of head trauma experimentally [7, 11, 16, 26].

The important role of free radicals in ischemic damage has forced scientists to use free-radical scavengers for therapy. It is known that endogenous enzymes such as glutathione peroxidase, superoxide dismutase and catalase all provide protection against free radical damage, and drugs such as alfa-tocopherol, polyethylene glycol-SOD and 21-aminosteroid tirilazad are still in the research area of restorative neurosurgery [33].

One of the effective antioxidant molecules has been shown to be the hormone melatonin. Both melatonin and methylprednisolone have been demonstrated to protect neurons, axons, myelin, and intracellular organelles including mitochondrion and nucleus. However, quantitative evidence has shown that this protection of neurons and subcellular organelles of spinal cord after secondary injury is much more obvious for melatonin than for methylprednisolone [25]. Two kinds of neuronal injury are defined for head trauma: One, the primary injury, which occurs at the time of impact and the other, a secondary and a progressive process. Free radicals, produced during oxidative reactions formed after trauma, have been suggested to be responsible for the mechanism of the secondary neuronal injury. The effects of melatonin during the early posttraumatic period seem to be hopeful [8].

An important neurotoxic versus neuroprotective molecule is nitric oxide. The constitutive and inducible nitric oxide synthase enzymes, their availability, their capacity to synthesise nitric oxide, and the concentrations of nitric oxide produced by either of them, are all the determinants of injury producing or neuroprotective effects of nitric oxide [22]. Calpains are cytosolic neutral cysteine proteases that are activated by increased intracellular calcium levels, and it has been demonstrated that calpain inhibitors are likely to play a role in the prevention of delayed injury resulting from axonotomy [57]

It is well known that inflammatory cascade worsens ischemic injury due to the toxic effects of inflammatory cytokines on endothelial cells. Thus, intercellular adhesion molecules, neurotrophic and growth factors (nerve growth factor-NGF, basic fibroblast growth factor-bFGF, ciliary neurotrophic factor, brain derived neurotrophic factor (BDNF), insulin-like growth factor, glial cell-derived neurotrophic (GDNF), neurotrophin-3, neurotrophin-4/5) are all under research for restoration of neuronal injury [28, 32]. Endothelin peptides are another group of endogenous substrates for neuronal injury. Endothelin is a strong vasoconstrictor mediator found in the endothelial cells and endothelins receptor antagonists are important agents in the treatment of neuronal injury secondary to cerebral vasospasm [47, 60, 53].

Cell Transplantation Into the Central Nervous System

Grafting into the brain has always been interesting for scientists since the beginning of the last century

[10]. Thus, cell transplantation has emerged as a promising approach for restoration of functions in the central nervous system. Most of the work has been performed in rodents, but there are a few studies in primates and humans. Neurons may be transplanted to the site of cell loss or into the target of cell loss and survive, migrate, vascularize, establish new synaptic connections, preserve neurotransmitter content and may function. The grafts generally integrate both anatomically and functionally with the host system. Grafted fetal neurons can retain the capacity to interconnect with appropriate target neurons with functional connections [38, 39]. Transplantation of embryonic tissue into the anterior chamber of eye has also been demonstrated to show synaptic connections after implantation as well [54].

Transplants of many types of embryonic neurons into favourable sites in an adult brain can make extensive connections with the host with axons growing considerable distances. Transplants of adult or even early postnatal brain are generally much less successful: Most of the neurons either do not survive or those few that survive do not generally grow axons [6, 5, 41]. Source of donor tissue and age of host at the time of transplantation are important factors for graft survival. In the studies on central nervous system transplantation, best survival has been achieved in fetal grafts. It is well known that embryonic neural tissue can survive after transplantation into the brains of immature or mature recipients [24].

It is clear that while embryonic neural tissue can survive for prolonged periods after transplantation into the brain, it nevertheless exists in an immunologically unstable state. Embryonic neural tissue can be transplanted to neonatal and adult brains, where it matures and integrates with the host brain. While allografts, and in some cases xeno-grafts, survive for prolonged periods, they are nevertheless always susceptible to immune rejection. It is proposed that by examining the conditions that precipitate such rejection, insight will be gained into the nature of immunological privilege as it pertains to the brain. Medawar in 1948 suggested that brain was the immunologically privileged site of the body and later indefinite xenograft survival and prolonged survival of allograft was reported by scientists [31, 35, 48]. Yoffey *et al.* reported that the brain has a limited lymphatic drainage and that graft-associated antigens are prevented from gaining access to the host brain. If the host brain is sensitized, graft rejection does occur [13, 58]. The presence of blood-brain barrier was considered an important barrier for the success of neural grafting as well [50]. Another important point has been suggested to be the absence of antigen-presenting (class 2-positive) cells in the brain [20].

Due to the limited availability of human donor material and ethical concerns with its use, porcine tissue has been considered an appropriate alternative. In animal studies, neural allo- and xenografts are usually rejected in the brain, emphasizing the necessity of understanding factors underlying the survival and rejection of intracerebral neural transplants. Rejection of neural xenografts is expected to be of a cellular nature, like neural allograft rejection, but may also display unique features and cannot be dealt with using conventional immunosuppressive therapies. The challenge therefore is to improve existing strategies and design new ones that allow permanent survival of histoincompatible neural grafts, taking advantage of the special immune status of adult CNS and immature donor brain tissue. The role of the immune system is further emphasized by the fact that rejection is prevented or slowed down by treatment with cyclosporine. It must be realized that such grafts are not, as is sometimes presumed (e.g. Madrazo *et al.*, 1987), stable and that graft rejection appears to be a somewhat capricious process [34].

The cytoprotective effect of iloprost was studied on isolated embryonic cortical brain tissue grafts of rats. It was demonstrated that iloprost significantly protected the neuronal integration capacity of the tissue pieces compared with saline preserved pieces. Tissues preserved in iloprost showed only minimal dissolution of the tissue with minimal extracellular edema only in the later stages of preservation. The cytoprotective effect of iloprost on the viability and survival of embryonic cortical brain tissue grafts was examined and integration of the graft tissue into the host brain tissue, a higher cellular population with new vascularization areas and preservation of myelin formation were accepted as a desirable, successful survival [42, 43].

Neural transplantation offers great hope for the treatment of progressive neurodegenerative disorders. Clinical trials have so far focused on the use of implants of embryonic mesencephalic tissue containing, already fate-committed, dopaminergic neuroblasts with the capacity to develop into fully mature dopamine neurons in their new location in the host brain. Restoration of function in neurodegenerative diseases, in particular Parkinson's and Huntington's disease, have been studied using transplantation of autologous adrenal medullary tissue for Parkinson's disease and human fetal transplants for Huntington's disease [2, 34].

Recent advances in stem cell biology, including methods of cell amplification and control of differentiation in vitro, represent her frontiers to neuroscientists because of the potential of stem cells for restoration of function of damaged brain and spinal cord. In recent years, it has become evident that the developing, and even the adult, mammalian CNS contains a population of undifferentiated, multipotent cell precursors, neural stem cells, which might be of advantage for the design of more effective restorative approaches for many neurological diseases [40]. Two different stem cells: embryonic and adult stem cells have been shown to have the capacity to differentiate into different cell types and have restricted potential for generating variable pathology [15, 49]. Stem cells are pluripotential and have the ability to proliferate and migrate. The most significant advances in the stem cell transplantation field will be the research strategies on predifferentiation of stem cells prior to transplantion, and factors affecting stem cell differentiation in the complex environment of the CNS in vivo [27].

Special Target of Restorative Neurosurgery

Parkinson's disease (PD) is an incurable neurodegenerative condition of the central nervous system (CNS) that typically presents in the fifth to seventh decade of life, with a movement disorder that consists of a resting tremor, bradykinesia and rigidity. As is known, medical therapy for Parkinson's disease (PD) often becomes inadequate over several years. Disability increases despite maximal medical management and many patients develop motor fluctuations and dyskinesia. In addition, medications provide good control of tremor in only 50% of cases. In appropriately selected cases, surgical therapies for PD provide benefit over medically refractory symptoms. Cell replacement for restoration of neurological functions in patients with movement disorders has been investigated for more than 15 years. Initial attempts used autologous adrenal medulla grafts implanted into the denervated striatum of patients with Parkinson's disease (PD). This approach was soon abandoned in favor of intrastriatal implantation of human embryonic mesencephalic tissue, rich in dopaminergic neurons. Several open-label research trials have shown clinically meaningful improvement in parkinsonian signs and symptoms after striatal transplantation of allogeneic fetal ventral mesencephalic tissue. However, ethical concerns, variability in surgical techniques, and reports of unusual late complications in some patients have limited the use of allogeneic fetal ventral mesencephalic tissue. Research with alternative cell sources such as porcine fetal ventral mesencephalic and allogeneic retinal pigment epithelial cells have shown promising results in preclinical trials, and they are currently being tested in clinical trials. Presently, the most important research strategy to improve the functional recovery after transplantation is to increase the survival of grafted DA neurons and the density and extent of the dopaminergic reinnervation in the striatum [3, 18, 55, 59].

Huntington's disease is a fatal neurological autosomal dominant disorder, associated with a trinucleotide repeat and characterized by chorea and deterioration in cognitive and neuropsychiatric function. Primary pathological changes are found in the striatum, where GABAergic neurons undergo degenerative changes. Local interneurons are relatively spared. Clinical trials of fetal striatal tissue transplantation for the treatment of Huntington's disease are hopeful [14]. In Alzheimer's disease the etiology is still unknown and patients have decreased levels of acetylcholine, which is an important neurotransmitter for cognitive functions. Fetal allograft transplantation into the caudate nucleus, gene therapy, neurotrophic infusions are the target of the research. Stroke remains a major brain disorder that often renders patients severely impaired and permanently disabled. There is no available treatment for reversing these deficits. Hippocampal, striatal and cortical grafting studies demonstrate that fetal cells/tissues, immortalized cells, and engineered cell lines are likely to survive in the ischemic adult brain, correct neurotransmitter release, establish both afferent and efferent connections with the host brain, and restore functional and cognitive deficits in specific models of stroke. The success of neural transplantation depends on several factors such as the stroke model (location, extent, and degree of infarction), the donor cell viability and survival at pre- and post-transplantation, and the surgical technique, among others. The first clinical trial of neural transplantation in stroke patients is a mile-stone in stroke therapy, but subsequent large-scale trials should be approached with caution. No direct treatment is recognized as safe and effective for reversing the stroke-induced brain damage and functional/cognitive deficits.

The therapeutic potential of neural stem cell injections to hypoxic-ischemic brain injury has been shown. It appeared that these cells migrated preferentially to the site of ischemia, experienced limited proliferation, and differentiated into neural cells lost due to injury, and reconstituted brain damaged by hypoxic-ischemic brain injury. Preliminary data in animal models of stroke lends support to these hypotheses [36, 46].

After brain injury progressive deterioration in cognitive and behavioral function and cholinergic neuronal degeneration may occur. Use of grafts of axonal growth substrates with nerve growth factor to promote morphological and functional recovery in the rat brain after lesioning of septohippocampal projections have been shown to be promising. Gene therapy has also been shown to modify cholinergic neuronal degeneration in the adult primate [56].

In restoration of spinal cord functions, there are two basic trends: Prevention of secondary changes and attempts for returning certain functions back. Because of the lack of regeneration in the spinal cord, further experiments such as transplantation procedures are requested. Central nervous system derived grafts have been used in spinal cord injury. As is known, descending monoaminergic tracts control locomotor and autonomous functions from locus cereleus and raphe nucleus [52]. After transplantation of noradrenergic neurons from embryonic brainstem in a transected spinal cord model, reinnervation, synapsis and locomotion have been recorded [17]. Transplantation of embryonic serotoninergic neurons from raphe nucleus into the spinal cord has also been shown to reinnervate, establish synapsis and functionality such as locomotion [12]. Central nervous system derived grafts are naturally occurring grafts, survive longer periods and have the capacity to course along both gray and white matter [37].

Dorsal root ganglion cell transplantation directly into the contused spinal cord, and dorsal root ganglion cells with Schwann cells transplantation directly into the contused spinal cord are suggested to have a potential of therapeutic value. They have potential of culturing from host peripheral nerve tissue and have the capacity to become vascularized [21, 30]. Olfactory sheat glia transplantation into the spinal cord has been reported to promote axonal regeneration as well [23]. Multipotent progenitor cell transplantation into the center of the lesion host axons have also been demonstrated to enter into the graft and induce some improvement in gait production [51].

Neurons in the central nervous system have a remarkable capacity to regenerate their transected axons when provided with an appropriate growth environment. Growth inhibitory proteins block axon regeneration in the central nervous system, and many of these proteins have been identified. Methods that stimulate regeneration in the spinal cord are antibodies that bind inhibitory proteins in myelin and allow axon regeneration in the central nervous system; methods that modulate neuronal intracellular signaling allow axons to grow directly on the inhibitory substrate of the central nervous system, and lastly transplantation of cells to the lesioned spinal cord promotes repair. Peripheric nervous system Schwann cells release a number of characterized and uncharacterized neurotrophic factors that exert powerful regeneration promoting influences on axons in the PNS. Thus it has been hypothesized that implantation of Schwann cells, or infusion of factors that release into the lesioned spinal cord, should lead to central nervous system regeneration [29].

The pineal gland is an endocrine organ which exerts regulatory effects on the activity of various organs and systems. Research about the pineal gland are ageing, locomotion, free radical scavenging effect, antitumoral effect, which are all under research for restorative neurosurgery [4, 44, 45].

Conclusion

In conclusion, great emphasis is currently given to restorative neurosurgery to promote survival from primary or secondary neuronal injury due to several neurodegenerative diseases, ischemic or traumatic CNS disorders. Developments both in neurobiology and biotechnology will be the most supporting aspects of restorative neuroscience and neurosurgery.

References

1. Attardi Dg, Sperry RW (1963) Preferential selection of central pathways by regenerating optic fibers. Exp Neurol 7: 46–64
2. Backlund EO, Granberg PO, Hamberger B, Gedvall G, Seiger A, Olsen L (1985) Transplantation of adrenal medullary tissue to striatum in Parkinsonism: first clinical trials. J Neurosurg 62: 169–173
3. Barker RA (2000) Prospects for the treatment of Parkinson's disease using neural grafts. Expert Opin Pharmacother 1: 889–902

4. Berker M, Palaoglu S, Ozcan OE, Erbengi A (1996) The effect of pinealectomy and foetal pineal transplantation of collagen ageing in rats. Acta Neurochir (Wien) 138: 1224–1228; discussion 1228–1229
5. Berkman MZ, Palaoglu S, Erbengi T, Erbengi A (1998) Neurotransmitter and amino acid analysis and ultrastructural observations of fetal brain cortex transplantation to adult rat brain under the effect of dexamethasone. Neurosurgery 42: 1126–1133; discussion 1133–1134
6. Björklund A, Stenevi U (1984) Intracerebral neural implants: neural replacement and reconstruction of damaged circuitries. Ann Rev Neurosci 7: 279–308
7. Choi DW (1995) Calcium: still center-stage in hypoxic-ischemic neuronal death. Trends Neurosci 18: 58–60
8. Cirak B, Rousan N, Kocak A, Palaoglu O, Palaoglu S, Kilic K (1999) Melatonin as a free radical scavenger in experimental head trauma. Pediatr Neurosurg 31: 298–301
9. David S, Aguayo AJ (1981) Axonal elongation into peripheral nervous system bridges after central nervous system injury in adult rats. Science 241: 931–933
10. Dunn EH (1917) Primary and secondary findings in a series of attempts to transplant cerebral cortex in the albino rat. J Comp Neurol 27: 565
11. Ercan M, Inci S, Kilinc K, Palaoglu S, Aypar U (2001) Nimodipine attenuates lipid peroxidation during the acute phase of head trauma in rats. Neurosurg Rev 24: 127–130
12. Foster GA, Roberts MH, Wilkinson LS, Björklund A, Gage FH, Hokfelt T, Schultzberg M, Sharp T (1989) Structural and functional analysis of raphe neurone implants into denervated rat spinal cord. Brain Res Bull 22: 131–137
13. Freed WJ (1983) Functional brain tissue transplantation: reversal of lesion-induced rotation by intraventricular substantia nigra and adrenal medulla grafts, with a note on intracranial retinal grafts. Biol Psychiatry 18: 1205–1267
14. Freeman TB, Hauser RA, Willing AE, Zigova T, Sanberg PR, Saporta S (2000) Transplantation of human fetal striatal tissue in Huntington's disease: rationale for clinical studies. Novartis Found Symp 231: 129–138; discussion 139–147
15. Gage FH (2000) Mammalian neural stem cells. Science 287: 1433–1438
16. Germano IM, Bartkowski HM, Cassel ME, Pitts LH (1987) The therapeutic value of nimodipine in experimental focal cerebral ischemia: Neurological outcome and histopathological findings. J Neurosurg 67: 81–87
17. Gimenez y RMG, Orsal D, Feraboli-Lohnerr D, Privat A (1998) Recovery of locomotion following transplantation of monoaminergic neurons in the spinal cord of paraplegic rats. Ann N Y Acad Sci 860: 393–411
18. Hagell P (2000) Restorative neurology in movement disorders. J Neurosci Nurs 32: 256–262
19. Hall SM (1986) The effect of inhibiting Schwann cell mitosis on the reinnervation of acellular autografts in the peripheral nervous system of the mouse. Neuropathol Appl Neurobiol 12: 401–414
20. Hart DNJ, Fabre JW (1981) Demonstration and characterization of la-positive dendritic cells in the inter-stitial connective tissues of rat heart and other tissues but not brain. J Exp Med 153: 347–361
21. Hoovler DW, Wrathall JR (1991) Implantation of neuronal suspensions into contusive injury sites in the adult rat spinal cord. Acta Neuropathol (Berl) 81: 303–311
22. Iadecola C, Zhang F, Casey R, Nagayama M, Ross ME (1997) Delayed reduction of ischemic brain injury and neurological deficits in mice lacking the inducible nitric oxide synthase gene. J Neurosci 17: 9157–9164
23. Imaizumi T, Lankford KL, Waxman SG, Greer CA, Kocsis JD (1998) Transplanted olfactory ensheating cells remyelinate and enhance axonal conduction in the demyelinated dorsal columns of the rat spinal cord. J Neurosci 18: 6176–6185
24. Jaeger CB, Lund RD (1980) Transplantation of embryonic occipital cortex to the brain of newborn rats. Exp Brain Res 40: 265–272
25. Kaptanoglu E, Tuncel M, Palaoglu S, Konan A, Demirpence E, Kilinc K (2000) Comparison of the effects of melatonin and methylprednisolone in experimental spinal cord injury. J Neurosurg 93: 77–84
26. Kaste M, Fogelholm R, Erila T, Palomaki H, Murros K, Rissan A, Sarna S (1994) A randomized, double-blind, placebo-controlled trial of nimodipine in acute ischemic hemispheric stroke. Stroke 25: 1348–1353
27. Keirstead HS (2001) Stem cell transplantation into the central nervous system and the control of differentiation. J Neurosci Res 1; 63: 233–236
28. Korfali E, Doygun M, Ulus IH, Rakunt C, Aksoy K (1988) Effects of neuronotrophic factors on adrenal medulla grafts implanted into adult rat brains. Neurosurgery 22: 994–998
29. Lance EM (1967) A functional and morphologic study of intracranial thyroid allografts in the dog. Surg Gynecol Obstet 125: 529–539
30. Kuhlengel KR, Bunge MB, Bunge RP, Burton H (1990) Implantation of cultured sensory neurons and Schwann cells into lesioned neonatal rat spinal cord: part II – implant characteristics and examination of corticospinal tract growth. J Comp Neurol 293: 74–91
31. Kuffler D (2000) Can regeneration be promoted within the spinal cord? P R Health Sci J 19: 241–252
32. Lindvall O, Ernfors P, Bengzon J, Kokaia Z, Smith ML, Siesjö BK, Persson H (1992) Differential regulation of mRNAs for nerve growth factor, brain-derived neurotrophic factor, and neurotrophin 3 in the adult rat brain following cerebral ischemia and hypoglycemic coma. Proc Natl Acad Sci USA 89: 648–652
33. Liu TH, Beckman JS, Freeman BA, Hogan EL, Hsu CY (1989) Polyethylene glycol-conjugated superoxide dismutase and catalase reduce ischemic brain injury. Am J Physiol 256: H589–H593
34. Madrazo I, Drucker-Colin R, Diaz V, Martinez-Mata J, Torres C, Becerril JJ (1987) Open microsurgical autograft of adrenal medulla to the right caudate nucleus of two patients with intractable Parkinson's disease. NEnglJMed 316: 831–834
35. Medawar PB (1948) Immunity to homologous grafted skin: II. The fate of skin homografts transplanted to the brain, to subcutaneous tissue, and to the anterior chamber of the eye. Br J Exp Pathol 29: 58–69
36. Nishino H, Borlongan CV (2000) Restoration of function by neural transplantation in the ischemic brain. Prog Brain Res 127: 461–476
37. Nornes H, Björklund A, Stenevi U (1983) reinnervation of the denervated adult spinal cord of rats by intraspinal transplants of embryonic brain stem neurons. Cell Tissue Res 230: 15–35
38. Olson L, Malmfors T (1970) Growth characteristics of adrenergic nerve in the adult rat – fluorescence histochemical and 3H-noradrenaline uptake studies using tissue transplantations to the anterior chamber of the eye. Acta Physiol Scand 348: 1–112
39. Olson L, Sieger A (1972) Brain tissue transplanted to the anterior chamber of the eye fluorescence histochemistry of immature catecholamine and 5-hydroxytryptamine neurons reinnervating the rat iris. Z Zellforsch 135: 175–194
40. Ourednik V, Ourednik J, Park KI, Teng YD, Aboody KA, Auguste KI, Taylor RM, Tate BA, Snyder EY (2000) Neural stem cells are uniquely suited for cell replacement and gene therapy in

the CNS. Novartis Found Symp 231: 242–262; discussion 262–269, 302–306
41. Palaoglu S, Benli K, Pamir N, Erbengi T, Erbengi A (1988) Examination of autologous and embryonic cortical brain tissue transplantation to adult brain cortex in rats. Surg Neurol 29: 183–190
42. Palaoglu S, Erbengi A, Sav A, Erbengi T, Turker RK (1990) Cytoprotective effect of iloprost on isolated cortical brain tissue grafts in rats. Prostaglandins Leukot Essent Fatty Acids 39: 203–206
43. Palaoglu S, Erbengi A, Kaya S, Erbengi T, Turker RK (1990) Improved survival of transplanted cortical brain tissue grafts by iloprost in rats. Surg Neurol 34: 212–218
44. Palaoglu S, Sungur A, Atasever A, Ruacan S, Akalin S, Ozgen T, Erbengi A (1994) Morphological assessment in pinealectomy and foetal pineal gland transplantation in rats: part I. Acta Neurochir (Wien) 128: 1–7
45. Palaoglu S, Palaoglu O, Akarsu ES, Ayhan IH, Ozgen T, Erbengi A (1994) Behavioural assessment of pinealectomy and foetal pineal gland transplantation in rats: part II. Acta Neurochir (Wien) 128: 8–12
46. Park KI (2000) Transplantation of neural stem cells: cellular & gene therapy for hypoxic-ischemic brain injury. Yonsei Med J 41: 825–835
47. Pluta RM, Boock RJ, Afshar JK, Clouse K, Bacic M, Ehrenreich H, Oldfield EH (1997) Source and cause of endothelin-1 release into cerebrospinal fluid after subarachnoid hemorrhage. J Neurosurg 87: 287–293
48. Raju S, Grogan JB (1977) Immunologic study of the brain as a privileged site. Transplant Proc 9: 1187–1191
49. Ramon-Cueto A, Cordero MI, Santos-Benito FF, Avila J (2000) Functional recovery of paraplegic rats and motor axon regeneration in their spinal cords by olfactory ensheating glia. Neuron 25: 425–435
50. Reese TS, Kamovsky MJ (1967) Fine structural localization of a blood-brain barrier to exogenous peroxidase. J Cell Biol 34: 207–217
51. Reier PJ, Stokes BT, Thompson FJ, Anderson DK (1992) Fetal cell grafts into resection and contusion/compression injuries of the rat and cat spinal cord. Exp Neurol 115: 177–188
52. Rossignol S, Chau C, Brustein E, Belanger M, Barbeau H, Drew T (1996) Locomotor capacities after complete and partial lesions of the spinal cord. Acta Neurobiol Exp (Warsz) 56: 449–463
53. Sato M, Noble LJ (1998) Involvement of the endothelin receptor subtype A in neuronal pathogenesis after traumatic brain injury. Brain Res 809: 39–49
54. Sieger A, Olson L (1977) Quantitation of fiber growth in transplanted central monoamine neurons. Cell Tissue Res 179: 285–316
55. Subramanian T (2001) Cell transplantation for the treatment of Parkinson's disease. Semin Neurol 21: 103–115
56. Tuszynski MH, U HS, Amaral DG, Gage FH (1996) Nerve growth factor infusion in the primate brain reduces lesion-induced cholinergic neuronal degeneration. J Neurosci 10: 3604–3614
57. Villa PG, Henzel WJ, Sensenbrenner M, Henderson CE, Pettmann B (1998) Calpain inhibitors, but not caspase inhibitors, prevent actin proteolysis and DNA fragmentation during apoptosis. J Cell Sci 111: 713–722
58. Yoffey JM, Courtice FC (1970) Lymphatics, lymph and the lymphomyeloid complex. Academic Press, New York
59. Zesiewicz TA, Hauser RA (2001) Neurosurgery for Parkinson's disease. Semin Neurol 21: 91–101
60. Zimmermann M, Seifert V (1998) Endothelin and subarachnoid hemorrhage: an overview. Neurosurgery 43: 863–875

Correspondence: S. Palaoğlu Hacettepe University, School of Medicine, Neurosurgery Department, Altındag, 06100 Ankara, Turkey.

Acta Neurochir (2002) [Suppl] 83: 101–104

Ethics in Publication

M. O. Guc

Department of Pharmacology, Faculty of Medicine, Hacettepe University, Ankara, Turkey

Summary

In this review, various facets of scientific communication are explored from the ethics point of view and specific questions related to the relevant steps of producing a scientific publication are addressed. Firstly, a brief overview of the ethics concept is presented with its association to other moral directives such as traditions, law and conscience, while the intersections of logic and faith with the hypothetic boundary of ethics, are analyzed by using a Venn diagram. Secondly, the everchanging concept of scientific (co)authorship is evaluated according to the degree of intellectual contribution to the final outcome and a brief emphasis is placed on the importance of an urgent need for rapidly developing the ethical rules for electronic publication in cyberspace. And lastly, the characteristics of different forms of scientific misconduct are summarized.

Keywords: Ethics; publication; misconduct; Venn analysis; coauthorship.

Science and Publication

In the quest for better understanding the "universe", scientists ought to share information with each other, critique the ideas, data or hypotheses, and pursue the opportunity of reaching a concensus with their peers on distinct issues. Scientific journals that publish this knowledge, provide a convenient medium for a significant proportion of this process. In a transforming world of growing emphasis to link the "scientific success" with "scientific publication", it is becoming inevitable to be encountered with variable grades of compromise in ethical integrity. Moreover, the unashamed anxiety caused by the "publish or perish" dictum also gets its share from the overall burden exerted on all parties concerned, i.e. the researchers, referees, editors and the public. The dilemma further intensifies with the solid fact that "science does not exist until it is published" [1].

Although, breaches from the ethical framework can be unanimously classified as a defiant injury to the achievements of humanity, there may be noteworthy differences among the subclasses of scientific disciplines with regard to the immediate impacts inflicted on the "society". This point can be clarified by choosing the examples from two extremes, such as socio-linguistics and medicine. It is easily acceptable that a similar degree of deviation from the "absolute scientific fidelity" in both disciplines will have profound differences in their instant effects on civilization. In socio-linguistics, the deleterious effects are expected to occur in a rather isolated environment with a considerable lag period compared to medicine, which receives unprecedented attraction from the public. Especially, if misleading happened to occur at medical operations, with the potential to be utilized on a global scale (e.g. misinformation on a new vaccine to be used widely), the suffering will undoubtedly be exemplified.

Ethics Concept

Before getting into the intricate details of ethics and publication relationship, it is useful to clarify the boundaries of ethics with regard to other co-existing sets of morale values. One way of addressing this is to perform a conceptual Venn analysis by using rather sound concepts such as laws and logic, in contrast to some conjectural notions, like conscience and faith. A brief outline of such an attempt is given in Fig. 1.

Although the suggested layout of these boundaries are far from being definite, it is possible to allocate the stand points of many perceptions related to the ethics.

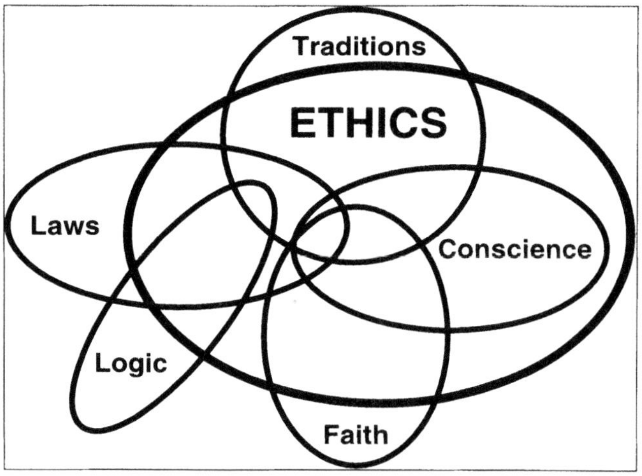

Fig. 1. Venn diagram representing some concepts related to ethics

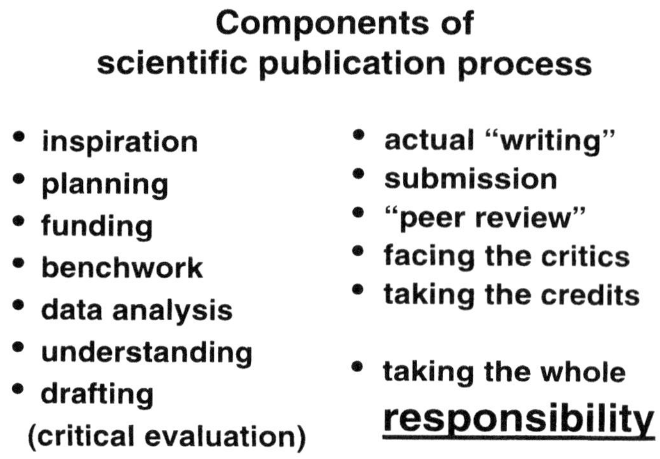

Fig. 2. The components of scientific publication process

One such item, which deserves special emphasis, appears to be conscience. According to the Venn diagram in Fig. 1, it is depicted to reside totally in the grand circle of ethics, implying that if an easy shortcut to remaining safely within the boundaries of "ethics" is needed, "conscience" may serve as the rule-of-thumb.

However, it should be emphasized that it is perfectly possible to stand within all the boundaries although some areas do not have intersections in the Venn diagram. In other words, an action can both be acceptable from the view of logic and faith while being entirely ethical and legitimate. Therefore, the Venn diagram in Fig. 1 should not be taken as the description of "different territories" but "related pathways" all amalgamating in the grand cauldron of moral values.

Scientific Author

The other party involved in scientific publication is *the author* whose description has evolved considerably in recent years. Who is a scientific author? A straight forward answer could be "someone who *writes* science"! Unfortunately, such an easy answer to this simple question is becoming more and more difficult to justify by the ever-changing process of scientific "writing". Since, nowadays, it is widely accepted that not necessarily one ought to literally *write* science to be tagged as an author, but should actively contribute to the process of creating a scientific publication. This transformation has taken place due to the fact that the production of science can seldomly be performed solely as reflected by the multiauthorship in the majority of the scientific publications for quite a long time. Therefore, the modern description of the scientific author is increasingly being hall-marked by the presence of "creativity". Thus, the current answer to the above-mentioned question has become "someone who *creates* science". This *de novo* raison d'être also finds its support from a Biblical quote, i.e. "the Creator has authored the Universe" [2].

Who Should Be Listed as a Co-Author?

A closer look to the components of a scientific publication process as outlined in Fig. 2 mandates that "to take full responsibility" is a prerequisite for any person to be listed as a co-author. In contrast, it should be emphasized that performing any single item listed on the left-hand column of Fig. 2, does not automatically give the person the right to qualify as a co-author.

A more structured approach to define the criteria of co-authorship can be found in the guidelines set by the Vancouver Group starting from 1978 [3].

Briefly, any person, (i) who is able and willing to defend the paper in public, (ii) who is confident about the integrity of the data contained in the publication, and (iii) who can take the full responsibility of the operations related to the publication can be eligible for a co-authorship [3].

On the contrary, anybody (i) who only collected the data, (ii) who only funded the study, (iii) who did not contribute intellectually or (iv) whose name is added to the publication for the sake of doing a favour (gift authorship) or for the purpose of getting undeserved extra credits from the scientific medium (guest/honorary authorship) does not qualify for such a co-authorship.

Besides these distinct subclasses of scientific authorship, the ethical rationalization of the ghost authorship still remains as a matter of debate. This is the situation usually when an innovative company operating in the scientific development area, hires manpower to "properly write" their findings for scientific journals. However, the trading of "scientific writing abilitiy" in return for currency as practiced in some non-English-speaking countries, attracts less ethical justification from the scientific league, despite some ambiguity.

Ethics in Cyberspace

Parallel to the expansive usage of internet in the daily life of the ordinary people, the incredible explosion of electronic communication among the scientists also urges the definition of the rules of ethical electronic publishing. Some attempts are being made in this area [4], and in the near future it is almost certain that we will witness the development of ethics in cyberspace.

Scientific Misconduct

The full understanding of ethics-science-publication relationship cannot be complete without discussing various facets of improper scientific behaviour. In its broadest definition, scientific misconduct is described as "any deviation from the norms of good scientific behaviour" [5] and it can be classified into two subgroups such as "intentional" and "unintentional" that have rather diverse specifications.

Unintentional Scientific Misconduct

Unintentional scientific misconduct bears no sinister hidden agenda as its name implies. This kind of misconduct may emerge because of (i) the improper qualification of the researcher that prevents him/her from accurately conducting the scientific operations, (ii) the lack of proper scientific research tradition within the institution that does not enable the scientist to receive a correct formation despite his/her demand, (iii) the lack of proper resources needed to conduct a healthy study, (iv) the lack of proper supervision from seniors and peers or (v) most frequently, because of the "sloppy research" that is due to the careless behaviour pattern in general. Whatever is the cause, unintentional type of misconduct can be alleviated by proper interventions.

Intentional Scientific Misconduct

The case of intentional scientific misconduct is much more serious to handle. In this type of ethical breach, there is always a pathological motive which sometimes deserves the intervention of parties other than the scientific community. The portfolio of this subclass starts with the (i) *abuse* of humans and animals involved in the research. The degree of abuse can vary from simple ignorance to serious violations of the basic rights as a living being.

Another group in this category is (ii) the *ignorantism* which is described as overlooking the rules, regulations and fundamentals of an accurate scientific behaviour. This type of misconduct may vary from intentional neglect towards the existing literature, to disregard the necessity that experiments should be repeated for a certain number of times in order to be both statistically and scientifically reliable.

(iii) *Plagerism*, that is described as the act of stealing the ideas of others or simply repeating what is already known, can be allocated to this subgroup of scientific misconduct together with (iv) the *salamization*. In the latter, the findings are intentionally divided into "least publishable units" which are then used to inflate the image of the researcher on that topic. The opposite of this behaviour (v) can be described as *"add-on"ization* which can be explained as follows. Usually the wrongdoer reports a batch of data for a start and waits with the second publication. When the time is right, the second batch of data is reported, amalgamated with the first batch, without making any discrimination between them. Therefore, the community gets the false impression that the data is more sound in a broad sense. Generally, it is believed that the act of add-onization lasts for a much longer time than anticipated, with no real stop.

One more type of scientific wrongdoing in this category (iv) can be termed as *data torturing*. It can be further classified into the "opportunistic type" in which the data are subjected to unnecessary but extensive statistical analyses until "it tells!" what the researcher wants, or into the "Procrustean type" named after the character from the Greek mythology. The legend tells that Procrust had lived near Athens and had a bed of a certain length beside him. Being a bully, he used to make the trespassing men or women lie in the bed and equalized the length of their legs exactly to the length of his bed, by either cutting or stretching as needed. Thus, in the "Procrustean type" of data torturing, the illicit researcher adjusts the results by selectively "culling" the outliers that do not look good in the data set.

A set of rather serious pathological behaviour patterns in this subclass comprise (v) *piracy* which is very similar to ordinary theft, in which the scientific crook coldheartedly steals the data, documents, papers, etc. of others and (vii) *fraud* which is an immoral act of deceit exerted by the person who uses any form of trickery and dishonesty. Additionally, (viii) *data fabrication* by performing "dry desk research" with no existing real data belongs to in this subclass of serious scientific misconduct.

It is obvious that the intentional type of scientific misconduct should be handled seriously by using all means of the scientific and non-scientific society; the unintentional type of scientific wrongdoer most probably is a candidate for a hopelessly lost case.

Therefore it is possible to conclude this assay on the ethical aspects of scientific publication by saying that "there appears to be nothing wrong about making mistakes in science provided that the conclusions are deduced critically, openly, and in good faith while the full responsibility is taken".

References

1. Davidoff F (2001) The other two cultures: how research and publishing can move forward together. In: Hudson Jones A, McLellan F (eds) Ethical issues in biomedical publication. The Johns Hopkins University Press, Baltimore, London
2. Hudson-Jones A (2001) Changing traditions of authorship. In: Hudson Jones A, McLellan F (eds) Ethical issues in biomedical publication. The Johns Hopkins University Press, Baltimore, London
3. International Committee of Medical Journal Editors (ICMJE) (1997) Uniform requirements for manuscripts submitted to biomedical journals. Annal Int Med 126: 36–47
4. McLelllan F (1998) Like hunger, like thirst: patients, journals and the internet. Lancet 352: 39–43
5. University of Michigan Medical School (1989) Guidelines for the responsible conduct of research. Ann Arbor, University of Michigan, USA

Correspondence: M. O. Guc, Department of Pharmacology, Faculty of Medicine, Hacettepe University, Ankara, Turkey.

Acta Neurochir (2002) [Suppl] 83: 105–108

The Art of Scientific Presentation

Ö. Palaoglu

Ankara University, Faculty of Medicine, Department of Medical Education and Informatics, Sıhhiye, Ankara, Turkey

Summary

While scientific research data can only become widespread by scientific communication, scientists generally underestimate the importance of communicating effectively. The present article is a current informational review of the guidelines for effective oral and poster presentation for young scientists. Emphasis has been given not only to effective structuring of the content but to effective verbal and non-verbal presentation techniques as well. References are mostly Internet sources, thus, have not been cited throughout the article but given at the end as a whole.

Keywords: Presentation skills; oral presentation; poster presentation; visual aids.

Introduction

Science, especially biological science is growing immensely, producing enormous amounts of new data. It is as a scientific research presentation that these new data are first introduced to the scientific community, to enable the colleagues to evaluate the observations and the outcome and repeat the experiments. Scientific research data become science only if shared among scientists and discussed on scientific platforms. However, the value of any scientific research data can fully be realised when it is effectively transferred to the target population.

Unfortunately, scientists who produce these enormous amounts of scientific data, generally underestimate the importance of presenting their data effectively, either orally or in written form. This is partly understandable since: effective presentation has generally been either perceived as a natural charisma or years of experience, and very little training has been received for effective presentation skills. However, effective presentation – whether it is for scientific research data or not – has its own principles; strategies and artistic nature have newly begun to be put on the agenda of scientists and have become an educational goal.

No need to discuss all generic strategies for an effective presentation which apply for scientific presentations as well. On the other hand, whatever the style, the main philosophy of scientific research presentation lies in the words *Its being simple that makes style, harmony, delicacy and good rhythm beautiful (Plato)*. It is this philosophy that makes scientific communication beautiful and enthusiastic but very difficult.

Scientific research data can mainly be presented as a full manuscript or as oral or poster presentation. The present article has been undertaken to review the basic principles of oral and poster presentation.

Oral Presentation

Oral presentations of scientific research data are generally 10, at most 20 minutes talk. Although there are many aspects of an effective oral presentation, in generic terms, the message, the speaker, and the audience are the critical ones to be taken into consideration first. Special attention must also be paid to how these critical components are treated while getting prepared for the presentation and during the presentation. Moreover, none of the components can be treated on their own but only together (Fig. 1).

The Message

Message is simply what the speaker says verbally and non-verbally, primarily including the content and the structure of the content. Although the cornerstone of an oral presentation is main data transfer, correct estimation of the nature of the target population, their level of knowledge, what they really know, what they don't know and what they would like to know enables the speaker to decide on and frame the content. One thing to be very optimistic is not to tell everything you

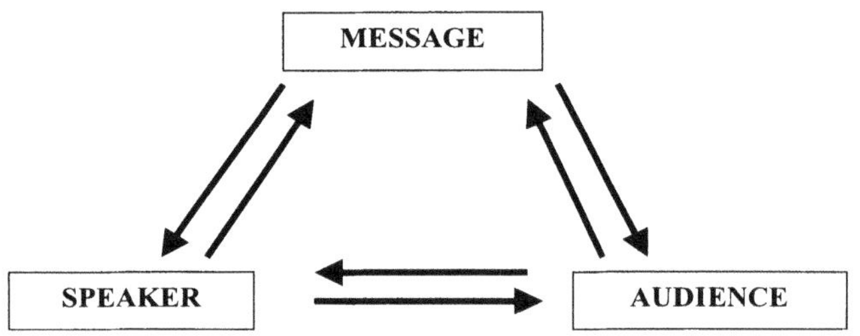

Fig. 1. The interaction between the major components of an oral presentation

know, but to tell only what the target population needs to know.

Organising the content may vary greatly. However, preference of a structure, offering an easy and familiar cognitive algorithm to scientists, is expected to make any scientific research data presentation more effective. The most familiar cognitive algorithm to any scientist is likely to be the algorithm used to solve scientific problems. IMRAD rule, which also covers the cognitive algorithm for scientific problem solving, is the most trustworthy systematic approach for structuring scientific research data. That is, answering *which*, *what* and *why* questions as the Introduction, *how it was done* question as Materials and Methods, *what was found* as Results and *what the results mean* as Discussion. Thus, a smooth and clear algorithm is of great importance because, if anytime during the presentation the audience loses attention and doesn't understand, there is no chance to go back and read.

Although universally accepted for all kinds of presentation, the power of each step can completely change according to the type of presentation and target population. Contrary to a written manuscript, it is preferable to give 70% of the allocated time of an oral presentation to main data transfer. For example, there is no need to describe the research methodology in detail, unless necessary, in an oral presentation: Anybody who wishes to know more can speak to you afterwards. Similarly, there is no need to mention all the background knowledge to define the problem and indicate its importance. Avoiding non-essential information is likely to help realising the central question being addressed as clear and simple. However, mentioning the approval of Ethical Committees must not be regarded as non-essential information and increases the credibility of the speaker. Although only 10% of the allocated time are proposed for discussion, the impact of the main message significantly increases with an explicitly clear last sentence.

Visual Aids

If correctly prepared and efficiently used, the visual aids are strongly complementary to oral presentations and very important reinforcers of the verbal communication. Although they allow the target population while listening to see the essential message of the presentation, they do not by themselves represent presentation, they are just aids for the speaker.

Visual aids generally used are 35 mm or computer slides and overhead transparencies. Although it is known that adults learn much more if different sorts of audio-visual aids are used, but since oral presentations of research data are generally short talks of 10–20 minutes, using more than one kind of visual aid inevitably distracts the fluency of the talk. Thus, it is preferable to choose one kind of visual aid. Most often, slides either 35 mm or computer-based are favoured. It must be remembered that overhead transparencies are effective only for classroom size areas.

Careful attention needs to be given to prepare effective 35 mm or computer slides. Effectiveness simply relies on two acronyms KISS and KILL: Keep It Short and Simple; Keep It Large and Legible.

TIPS on Effective Slides

- Let each slide contain one message. Don't display more than the argument needs.
- Let the slides be reader friendly. This includes bold and clear lettering.

1. The slides must be free of grammatical and spelling errors.
2. Horizontally prepared slides are more effective.
3. Try to keep the size, font and colour consistent.
4. Use large fonts: 18–24 type size lettering minimum. Let the headings be 20% bigger than the text. Practically, if the slide is readable when held at arm's length in front of you, it will be readable by the target population as well.
5. Avoid using capital lettering all through the slide.
6. Use bold type so it will be easier to read.
7. For a text, don't use more than 5 words to a line; 6–7 lines on a slide and 35 words on a slide.
8. Keep the background simple to avoid visual overcrowding.
9. Practically prefer light and bright lettering to dark background. Yellow lettering over dark blue background works best. Try to avoid using more than 2–3 colours. (Fig. 2)

Red lettering
Red and blue together
Red and green together.
Brown and gray together.
Red, gold, bright blue and green as background

Fig. 2. Specific coloring restrictions on a slide

- For illustrations, prefer simple graphs and charts. Tables, especially if too detailed, are not effective. If possible, avoid using more than 3–4 columns and 5–6 rows. Pie graphs are well suited for showing relative parts of the whole, bar graphs for comparing magnitudes and line graphs for showing correlation or trends.
- Do not confuse the target population with too many slides. It is difficult to designate a certain number of slides to a 10–20 minute talk but enough time must be given to the target population to view each slide.
- If you are going to use the same slide more than once, always duplicate it. Avoid planning a backward sequence.
- Let the first slide contain the title of the presentation and the speaker's name(s) and affiliation(s).
- Rehearse the presentation absolutely with the visual aids.

Transparencies, as complementary visual aids, are most suited to classroom areas and if the presentation has to be given in daylight. Being simple and short but large and legible applies to effective transparencies as well. To be effective, at most 5–8 lines of text, specifically prepared graphics and diagrams as well as a frame are the basic points to be considered when establishing a transparency. During the presentation, position the overhead to the right if the speaker is right handed and to the left if left handed. This enables the speaker to face the audience easily and take notes if necessary.

The Speaker

The target population excellently senses any speaker who is enthusiastic, well-prepared, confident, proud but eager to share information. There are various methods of any speech delivery: The speaker can read a manuscript, can memorise what he is going to say, he can perform an impromptu speech or an extemporaneous one. To encourage a conversational quality, extemporaneous delivery has to be preferred and performed. Extemporaneous speaking involves a carefully planned and structured but not a written or memorised speech. Being well-prepared and well-rehearsed enables the speaker to be flexible to adapt the message to the target population, respond to their reactions and to conceptualise the dynamics of the speech.

- Face the audience; Maintain eye contact; Never speak towards the screen.
- Keep your body open to the audience; don't fold your arms or put your hands in your pockets
- Don't lean on the podium.
- Dress appropriately
- Be relaxed
- Never be arrogant
- Never read either the whole speech or the visual aids.
- Stay within time limits
- Thank the audience

Fig. 3. Tips to be an effective speaker

The language used needs to be concrete and free of technical jargon. Speaking slowly but loudly enough and using straightforward simple sentences are important to sound confident and competent.

When rehearsed well enough, starting off strong and with confidence, and having planned the first sentence, not only inhibits the anxiety of public speaking but motivates the audience as well. Keeping the audience interested throughout the entire presentation also needs special care. Instead of speaking in a monotone voice, varying the tone of voice, including appropriate emotion to the topic, pacing the rate of speaking according to the familiarity of the subject are all helpful to emphasise key points. Mumbling always makes the audience lose attention. Throughout the organisational pattern of the speech, the speaker needs to plan appropriate transitions from one point to the other. Words such as *also*, *because*, *however* and phrases like *on the other hand*, *for example and in other words* are most useful for transition.

Non-verbal gestures, facial expressions and body language during the presentation are important to attract and maintain attention of the audience as well. Standing, walking or moving about with appropriate non-verbal gestures are all preferred to sitting down and standing still with head down and without having eye contact with the audience (Fig. 3).

An effective speaker needs to estimate correctly the benefits of the visual aids and decide on how they will affect attention and memory to make an efficient visual-verbal link. Besides having prepared the visual aids properly, special care must be given during their presentation. To give the audience a moment to become oriented with each visual aid, in other words pausing before continuing helps to strengthen the verbal-visual link. Using the pointers only where necessary also helps the audience to focus on the main important points.

Speakers must be very careful to remain within the time limits reserved for them. If the speaker is able to leave the audience with a sense of completion, no need to say, he/she has performed an effective oral presentation.

The Audience

Understanding the audience is the third cornerstone of an effective presentation. Knowing the target population, the extent of their knowledge about the topic going to be addressed, their educational background and what they are expecting to learn are important to decide on the content, the structure and the style of the presentation. The speaker must be able to take into consideration the self-actualisation, to a lesser extent the self-esteem and psychological needs of the listeners. Analysing the audience carefully is very helpful not only to attract and maintain the attention of the audience but to getting adjusted and adapted to situations during the presentation as well.

Poster Presentation

Poster presentation has become one of the most preferred scientific communication style nowadays. It is possible to reach a larger audience with greater opportunity for discussion and feedback.

All the general rules for preparing a presentation apply to a poster presentation as well. Thus, IMRAD is the most preferred structural organisation for the cognitive algorithm; the difference being the power of each component of IMRAD. Although clearly and simply expressed, main data transfer is essential; the possibility of displaying more than one main message is an advantage of poster over oral presentations. The Introduction and Materials and Methods sections need not be very detailed. However, a brief summary of the results and statements of major conclusions are very important for a poster presentation to be effective.

Poster presentation is not only a presentation style but a visual aid as well. Thus, the more it allows the eye to travel a natural pathway, either down the columns or along the rows, the more reader friendly it is.

The *Title* of a Poster Presentation needs to be descriptive and short but attractive. It must be legible from 2.5–3 m away. In other words, font-size 30–36 should be preferred for lettering. The *text* must be legible from 1.5 m away with 20–24 font-size lettering. *Text and tables* (diagrams, graphs) should be well balanced with enough plain space, which all contribute to make the poster effective. If attention is going to be focused on more than one important point, the different issues need to be effectively illustrated, probably by short and descriptive *subtitles*. Graphs and diagrams provide a clearer statement of research results than do tables and text. Framing the question to be addressed and the conclusions enable the readers to quickly realise the problem. Despite all the generic guidelines for an effective poster presentation, the only limitation of data transfer for poster presentation is the creativity of the scientist.

In conclusion, effective oral or poster presentation of scientific research data is not essentially a natural charisma but just a skill with its own cognitive, psychomotor and communicative components which can be learned and experienced. It will only be regarded as *Art* if performed well.

References

1. Day RA (1988) How to write and publish a scientific paper, 3rd edn. Phoenix Oryx Press
2. Essex-Lopresti M (1980) Illuminating an address: a guide for speakers at Medical meetings. Med Edu 14(1): 8–11
3. Daffner RH (2000) Opinion: on improvement of scientific presentations. Am J Roentgenology 174(5): 1229–1231
4. Presentation Guidelines, http://www.common.org/conferences
5. Presenting a Professional Paper At A Scientific Meeting, Tips and Guidelines, http://www.pantechnica.com
6. Preparation of Visual Aids For A Professional Presentation, http://www.pantechnica.com
7. Laskowski L, Elements of an effective speech, http://www.ljlseminars.com
8. Laskowski L, 7 aspects of a dynamic presentation, http://www.ljlseminars.com
9. Laskowski L, Using overhead transparencies, http://www.ljlseminars.com
10. Laskowski L, Using visual aids as notes, http://www.ljlseminars.com
11. Laskowski L, Impromptu or extemporaneous speaking, http://www.ljlseminars.com
12. Louderback AL, Abstract accepted: now what? a primer on slide preparation, http://www.the-scientist.com
13. Guidelines for the preparation of effective presentations, http://www.asp.org

Correspondence: Ö. Palaoglu, Ankara University, Faculty of Medicine, Department of Medical Education and Informatics, Shhiye, 06100 Ankara, Turkey.

Acta Neurochir (2002) [Suppl] 83: 109–113

How to Write an Experimental Research Paper

M. N. Pamir

Marmara University Institute of Neurological Sciences, Istanbul, Turkey

Summary

The art and practice of academic neurosurgery are mastered by defining and learning the pertinent basic principles and skills. This article aims to present general guidelines to one of the many roles of a neurosurgeon: Writing an experimental research paper.

Every research report must use the "IMRAD formula: introduction, methods, results and discussion". After the IMRAD is finished, abstract should be written and the title should be "created". Your abstract should answer these questions: "Why did you start?, what did you do?, what answer did you get?, and what does it mean?". Title of the research paper should be short enough to catch glance and memory of the reader and be long enough to give the essential information of what the paper is about.

Writing about the results of the experiment is no easier than the research itself. As surgery, writing a scientific paper is also an improvisation, but general principles should be learned and used in practice. The most effective style of learning basic skills to construct a research paper is the "trial and error" type.

Keywords: Neurosurgical education; research; experiment; abstract.

The Many Roles of an Academic Neurosurgeon (Fig. 1)

The art and practice of academic surgery are mastered by defining and learning the pertinent basic principles and skills, and by practicing them under an experienced mentor. This requires a formation of personal qualities and professional skills needed to integrate surgical practice and investigation, and their delineation through an appropriately structured training program.

In the words of C. M. Balch [2]: "The academic surgeon must exhibit the following professional traits:

- Be an excellent clinical neurosurgeon
- Be willing to initiate and test clinically relevant ideas
- Be a good communicator who willingly passes knowledge to others in an understandable way
- Be highly organized and able to maintain an appropriate balance among professional and personal objectives
- Have excellent interactive skills to facilitate research collaboration and fulfillment of administrative duties".

As the diversity and multiplicity of challenges confronting the academic surgeon have become so great that we can consider the pentathlete as an analogy. The pentathlete competes in multiple areas-swimming, fencing, shooting, running and riding-but is not record holder in events that are the single focus of other athletes.

When the "pentathlete" skills required of academic neurosurgeons become so complex that combining them in a single individual may compromise each skill too much, a reasonable substitute is awareness and appreciation for the manner in which each of these skills contributes to the whole. Nevertheless, lifelong neurosurgical education has to teach, develop, perfect and maintain several areas of professional and personal skills.

As an investigator the academic neurosurgeon has to learn how to initiate and test reasonable ideas, to develop basic laboratory experience and skills, and to write and publish the findings of the research. This article concentrates on one of the diverse academic traits that a neurosurgeon has to develop: How to write an experimental research paper.

Writing an Experimental Research Paper

Writing of the results of the experiment is no easier than the research itself. A. V. Pollock [8] clarifies that

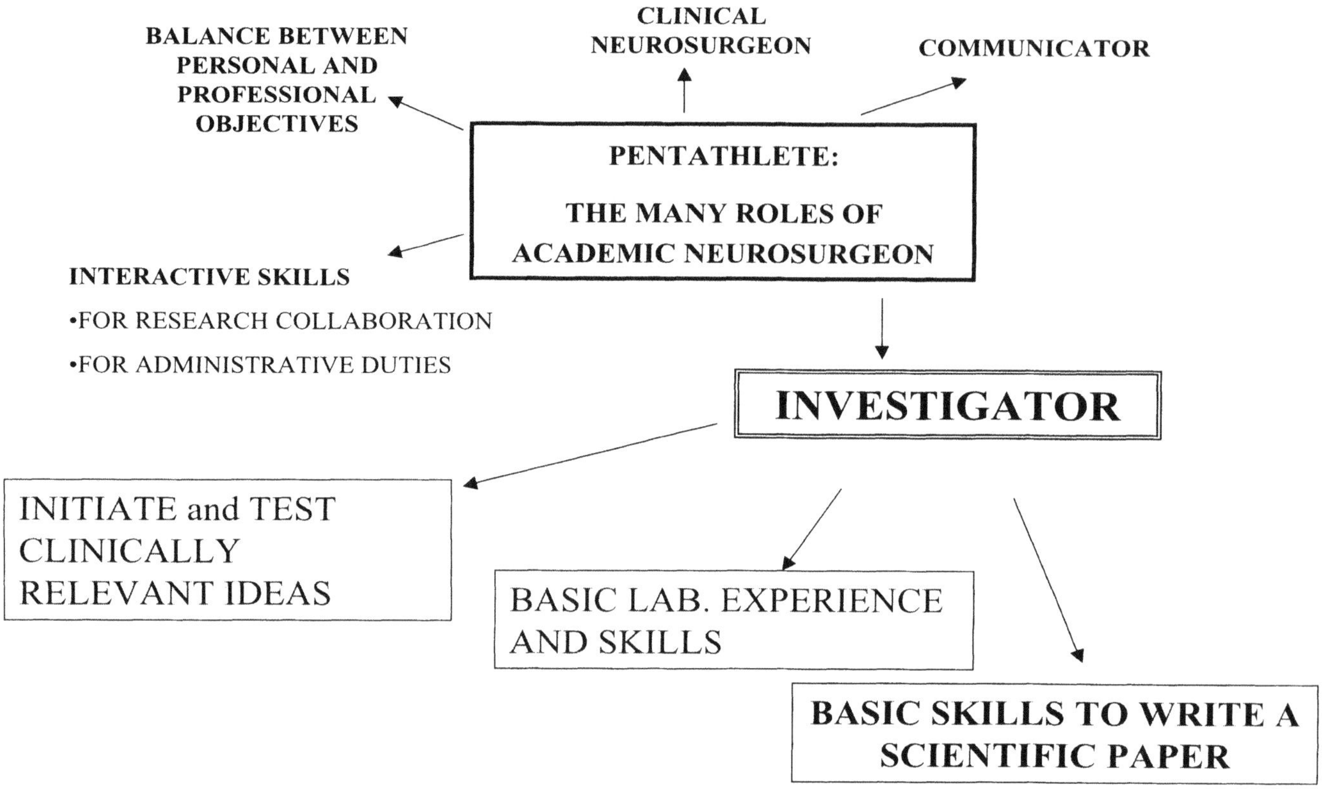

Fig. 1. This diagram summarizes the required skills and traits of an academic neurosurgeon with the "pentathlete analogy"

anybody who suggests that writing a scientific paper is easy has never written one. Nevertheless, the gift of writing is not something given to some people and not to others, but is an academical trait that can be taught and learned.

The main principle is to tell the truth as you find it in a clear way; moreover one has to try writing in such a way that it will be impossible to be misunderstood. After gathering enough reliable data to test your hypothesis, as an initial step read the "instructions of authors" of the journal to which you intend sending the manuscript. (In 1997, the International Committee of Medical Journal Editors has reissued its Uniform Requirements for Manuscripts Submitted to Biomedical Journals, summarizing the criteria of over 500 journals [7].)

"IMRAD" Formula

When you report your research you must use the "IMRAD" formula: introduction, methods, results, and discussion. When you have finished "IMRAD", you must write an abstract and give the paper a title. Because the abstract always comes before the introduction in a scientific paper, it will be considered first, but you will write it only after you have finished writing and editing the text.

Abstract

An abstract of your scientific paper is different than the abstract sent for a congress presentation. Whatever the format, be sure that the whole message of your work comes across in the abstract, because the part of the research paper that is most likely to be read is the abstract. Computer databases such as PubMed, Medline publish only abstracts. So, vague or incomplete abstracts will be overlooked by the other investigators. This may be one of the reasons why only about half of all published papers are ever cited.

In 1965 Bradford Hill [5] determined the questions that every abstract should answer:

1. Why did you start? (Introduction)
2. What did you do? (Methods)
3. What answer did you get? (Results)
4. What does it mean? (Discussion)

Initiated by the ever-increasing number of papers published each year (over two million in about 20.000 journals), a working group was set up at McMaster University in Canada to improve the quality of the information in abstracts [1]. The idea of "Structured Abstracts" was introduced and supported. In 1987, EJ Hugh, [6] the editor of Annals of Internal Medicine first used the format of structured abstracts.

Acta Neurochirurgica determines the components of the abstract as background, method, findings and interpretation. *Journal of Neurosurgery* divides it as objective, method and conclusion, and *Neurosurgery* structures the abstracts as objective, method, results, and conclusion. In *Cancer Research* e.g. there is no determined structured abstract format. Whatever the format is, structured abstract format helps the author present the whole message of the work in a planned way.

In structured abstracts each heading should start a new paragraph, all numbers should be written as numerals, abbreviations may be used if they are spelled out first and references should not be given. The length of the abstract is determined by the journal, but it must not exceed 400 words max, because it is the limit set by MEDLINE.

Components of an Abstract

Introduction or objective: This part should give, usually in one sentence, a precise statement of why the study was done. The hypothesis tested should be stated clearly.

Design: Type of the study, e.g. 'double blind trial', 'experimental analysis', should be noted.

Setting: To outline the conditions of the setting is more important for clinical research papers, so that the readers can assess the applicability of the study to their own circumstances. This paragraph should state whether the setting was the community, a university department etc.

Subjects or Material: The total number of patients, subjects or animals should be given together with a note of how they were selected. This will give the reader an idea of generalizability of the results.

Interventions: This part should include a description of any intervention e.g. the dosage and duration of a drug regimen.

Main Outcome Measures: Methods by which patients were assessed or the success of experiments judged should be mentioned.

Results: The main results should be given, together with a note of exclusions and withdrawals. Confidence intervals and level of significance should be indicated.

Conclusions: Only those conclusions supported by the data that are presented should be stated, followed by a short statement on the possible clinical applications of the work, bearing in mind the limitations of the study.

Title

The title of the research paper should be short enough to catch glance and memory of the reader and be long enough to give essential information of what the paper is about.

For example, the title of "Intracranial Inhibition of PDGF-mediated Glioblastoma Cell Growth by an Orally Active Kinase Inhibitor of the 2-phenylamino-pyrimidine Class: An Experimental Study" tries to give too much information in one sentence, and it is difficult to keep it in the short term memory. On the other hand, another form of this title as "Inhibition Glioblastoma Cell Growth: An Experimental Animal Study" is too short to give minimally required information.

Certainly the journal that the paper is intended to be sent to is another factor in determining the title, but one good style for the above example may be "Intracranial Inhibition of PDGF-mediated Glioblastoma Cell Growth by an Orally Active Kinase Inhibitor". For a neurosurgeon, this title indicates that this paper reports an experiment which studied the possible therapeutic role of an orally active molecule in the intracranial inhibition of glioblastoma cells.

Introduction

The introduction must set what you aimed to achieve and why you did pursue the goal in that particular way. A research paper is not a thesis, so it should not be a review of the literature although a few sentences about the historical perspective may be suitable. But most important, the hypothesis tested should be clearly stated.

Methods

The methods section can be an edited version of the research protocol in the beginning. You must be sure that the description of the methods that you used (in-

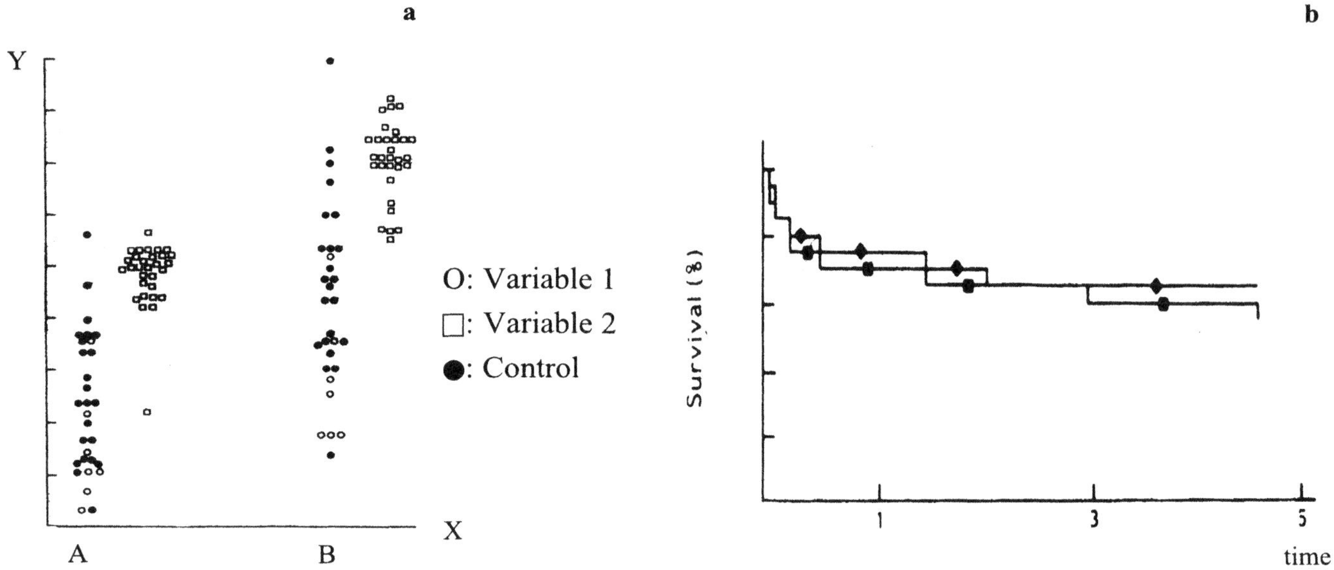

Fig. 2. There are two absolute indications for a graph rather than a table: data suitable for presentation as a scatterogram (a) and survival data (b)

cluding the statistical tests) is clear enough so that the work can be replicated.

Every neurosurgeon should be aware of the basics of biostatistics e.g. at least must learn the tests used for non-parametric versus parametric variables, but should always consult a biostatistician.

Results

The most important section of an experimental research paper is the results part. The investigator usually intends to present all the data that is generated with his/her great efforts. Only the necessary data with maximum simplification should be presented. In the results section, tables or graphs help the author simplify the information.

There are two absolute indications for a graph rather than a table: data suitable for presentation as a scatterogram and survival data (Fig. 2). Scatterograms compare multiple variables with regard to same parameter, and survival curves compare different variables, both types of information cannot be transferred in plain text only.

E. R. Tufte [9], in the book named as "The Visual Display of Quantitative Information" summarizes the rules of proper graph preparation in order:

- Show only the necessary data
- Maximize data-to-ink ratio
- Erase non-data ink
- Erase redundant ink data
- Revise
- Edit

"Data ink" describes ink on a graph that cannot be erased without losing the meaning of graph and non-data ink can be erased without loss of information (Fig. 3).

Tables must not repeat results that you have already detailed in the text or graphs. Each table must stand on its own, must not require reference to the text to explain it, and must not contain abbreviations. Each table must disclose whole numbers with or without percentages; percentages alone are not acceptable. Just as the remaining manuscript, the tables must also be typed double spaced.

Discussion

The discussion must be done depending on the data that you presented in the results. The author must obey the rules of composition at all times but additionally he must try to impart personality into the discussion, but he must never speculate, or contemplate the data. One should avoid too much use of "may" and "might". The conclusions drawn should be always supported by the data presented [3].

Modest conclusions may be more effective than bold

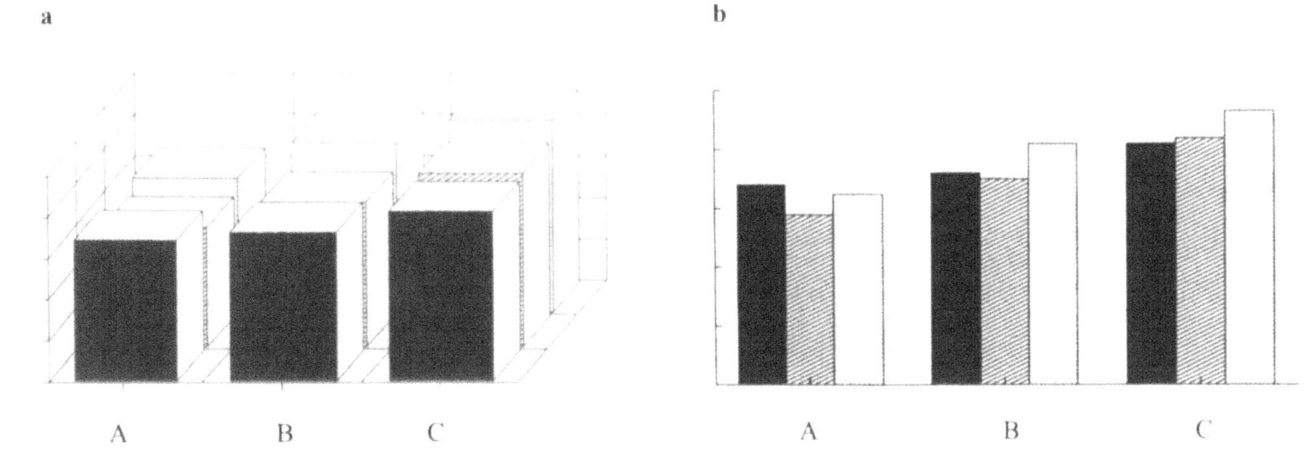

Fig. 3. "Data ink" describes ink on a graph that cannot be erased without losing the meaning of graph and non-data ink can be erased without loss of information. So data/ink ratio should be kept at maximum. Compare Figs. a and b, note that Fig. a displays non-data ink, which should be erased as much as possible as in Fig. b

claims: For example, in 1929, Alexander Fleming [4] reported in British Journal of Experimental Pathology, "Penicillin, in regard to infections with sensitive microbes, *appears to have* some advantages over the well-known antiseptics." Watson and Crick [10] reported in 1953: "It has not escaped our notice that the specific pairing we have postulated *suggests a possible* copying mechanism for the genetic material".

Authors must be concise in the discussion, editors will not give credit for a discussion that is twice as long as the rest of paper. Another common pitfall is to repeat the results, one must not repeat the results but comment on them in the light of substantial contributions of other researchers.

References

Authors must be selective in determining the references. One common mistake is to mention every paper that literature search has identified. Only the references really helpful in the research should be cited.

Conclusion

Although in residency many rules are being taught as surgical principles, improvisation is the main style of surgical art and science (and life, as well). Writing a scientific paper is also an improvisation like surgery, but after assimilating the main principles stated in the article.

The most effective mode of learning basic skills to write a scientific paper is the "Trial and Error" type. Comments of the editorial reviewers are the best educators. In the end, every pentathlete falls once.

References

1. Ad Hoc Working Group for Clinical Appraisal of the Medical Literature (1987) A proposal for more informative abstracts of clinical articles. Ann Intern Med 106: 598–604
2. Balch C, McKneally M, McPeek B, Mulder D, Spitzer W, Troidl H, Wechsler A (1997) Pentathlete: the many roles of academic surgeon. In: Troidl H, McKneally M, Mulder D, Wechsler A, McPeek B, Spitzer W (eds) Surgical research. New York: Springer, Berlin Heidelberg New York Tokyo
3. Condon R (1986) Type III Error. Arch Surg 121: 877–878
4. Fleming A (1929) On the antimicrobial action of cultures of a penicillium, with special reference to their use in the isolation of B. influenzae. Br J Exper Pathol 10: 226–236
5. Hill A (1965) The reasons for writing. Br Med J 2: 626–627
6. Hugh E (1987) Structured abstracts for papers reporting clinical articles. Ann Intern Med 106: 626–627
7. International Committee of Medical Journal Editors (1997) Uniform requirements for manuscripts submitted to biomedical journals. New Engl J Med 336: 309–315
8. Pollock A, Evans M (1997) Writing a scientific paper. In: Troidl H, McKneally M, Mulder D, Wechsler A, McPeek B, Spitzer W (eds) Surgical research. Springer, Berlin Heidelberg New York Tokyo
9. Tufte E (1983) The visual display of quantitative information. CT: Graphics, Cheshire
10. Watson J, Crick F (1953) Molecular structure of nucleic acids. A structure for deoxyribose nucleic acid. Nature 4356: 737–738

Correspondence: M. Necmettin Pamir M.D., Marmara Universitesi Norolojik Bilimler Enstitusu, PK 53 Maltepe, Istanbul, Turkey.

Acta Neurochir (2002) [Suppl] 83: 115–120

The Structure of a Neurosurgical Manuscript

N. Altınörs

Department of Neurosurgery, Başkent University, Ankara, Turkey

Summary

Preparing publishable manuscripts is an important aspect of professional life in research or clinical neurosurgery, especially for those in academic circles. Scientific writing skills can be developed through a long process of training and experience, starting with the residency program. The first step in developing a manuscript is to focus on a subject or problem that might be of significant interest to colleagues in the field. Next, the prospective writer must do a detailed survey of the relevant literature, the results of which will help him or her decide whether to actually write about the topic. Since the primary goal is to get the manuscript published, the writer should bear a specific journal in mind and write in accordance with the guidelines of that publication. He or she must also consider general ethics and scientific rules during the writing process.

Learning how to assess and use scientific sources, how to relate the collected information to the manuscript, and how to write in good scientific form are all important aspects of neurosurgical training.

Keywords: Biomedical journal; biomedical publication; information; neurosurgery.

Introduction

The first question that should be asked in a discussion of how to write a manuscript in neurosurgery is, "Is a manuscript in neurosurgery any different from a manuscript in any field of medicine?" The answer is no. Then comes the second question, "Why do people write?" The answer to this can be found in two sayings: One is the old Latin proverb "*verba volent, scripta manent*," which means "words disappear while writing remains." The second and more common saying is "publish or perish." A person might also ask the more specific question, "Why do people in medicine write?" The sophisticated answer is that people in the medical field want to expand horizons by exchanging new discoveries, or by presenting different and beneficial applications of things that are already known. On the other hand, this question could be answered from the individual's perspective. For most authors, scientific writing is an avenue to achieving personal goals of career enhancement and prestige.

Material and Method

We have touched on some of the "whys" regarding this type of writing, but the next important issue is "how" to write a scientific manuscript. Billing [6] set down the basic principles of writing 120 years ago with the advice, "First have something to say, say it as briefly as possible, and stop when you have said it." These words still ring true today, and Billing's suggestions can be used to form an understanding of how to prepare a manuscript.

The initial step in this process is to develop an idea; a subject or problem that one believes would be interesting and worthwhile to write about. Good knowledge of the relevant literature is necessary in order to judge the scientific originality of any specific subject, so a detailed literature search is mandatory. With this knowledge in hand, it is easier to decide whether to go ahead and write about the idea.

If the writer chooses to proceed, background information can be found in traditional sources such as books, reviews, abstracts, journal articles, audiovisual aids, and personal communication. The advent of electronic communication has radically expanded the possibilities for information gathering. Today, there is a tremendous volume of information available, access to it is cheap and easy to obtain, and facts can be transmitted and exchanged over long distances very rapidly. In scientific writing, the main applications of electronic communication are in bibliographical searches, electronic journals (e-journals), and the wide variety of databases on the Internet [3, 14, 18]. Electronic media have made it possible to access recent literature in particular, but the same does not hold true for older published information. The paper by Wilkins [24] is a very helpful source for approaches to finding references written before 1960.

Once the data is gathered, the difficult task of the actual writing begins. For most authors, the writing stage involves a lot of procrastination. Computers make the practical aspects of the job easier, because it is possible to store pieces of information in accessible form and pull them out as needed during progress. The actual writing process depends in part on the author's personal habits. Everyone has his or her preferred style and order of writing, but the final form of the manuscript must conform to the classical "IMRAD" format required by the vast majority of biomedical journals [16]. This arrangement designates four main sections, namely, Introduction,

Material and Methods, Results, and Discussion, each with specific functions and content. In addition to its main portion, a full manuscript will also include a Title Page, Abstract and Key Words, References, and Acknowledgements. Each of these sections is discussed below in detail, in the order they should appear in a manuscript.

Title Page

This page contains the title of the article, which should be short, descriptive, informative and catchy. The authors' names, highest academic degrees, and institutional affiliation are also noted, as are the names of the department(s) and institution(s) where the work was done. The addresses of the corresponding author and of the author responsible for reprints should also appear on this page.

Authorship and the number of the authors is an important issue. Only those individuals who have actively participated in planning and conducting the work should be listed as authors. Every author shares equal responsibility for the scientific content of the manuscript. Authors are also responsible for protecting patient privacy, and for obtaining permission from the original author if diagrams, photographs, illustrations, or long quotations from other sources are used. They are also responsible for arranging copyright transfer. Honorary or gift authorship, although practiced by some, is not generally approved.

The International Committee of Medical Journal Editors (ICMJE) has specified the following criteria for authorship:

i) Conception and design, or analysis and interpretation of data,
ii) Drafting the article or revising it critically for important intellectual content,
iii) Final approval of the version to be published [5, 15].

Hoen *et al.* questioned how widely these criteria are known and used [13]. To investigate this, they sent a questionnaire to 450 authors of 115 original articles that had appeared in the Dutch Journal of Medicine. The results showed that 63.6% of the authors fulfilled the ICMJE criteria, although 59.8% were unfamiliar with those criteria. The co-authors' answers to the questionnaire showed that 21% of them did not fulfill the criteria. The conclusion drawn from the study was that the ICMJE criteria are not sufficiently known, but many authors apply them implicitly. Hoen and coworkers stated that more effort should be made to increase the knowledge and use of these criteria [13].

Bhopal *et al.* [5] carried out a similar study of researchers in England. They found that 76% of respondents supported the criteria for authorship, but few knew and used them. Sixty-two percent of the surveyed group disagreed with the concept that all three of the above-mentioned ICMJE criteria must be met for authorship. Furthermore, the opinions of journal editors and researchers differed regarding authorship criteria. Researchers placed a higher value on practical research contributions than did editors.

Supplying financial or material support and technical aid is not sufficient for one to be considered an author. Such contributions from individuals or institutions are acknowledged either on the title page or after the main text.

Durack [7] noted that more than 98% of articles published in the Boston Medical and Surgical Journal a century ago were written by single authors. Today, fewer than 5% of the papers in the New England Journal of Medicine are written by one person. In 1984, Freisinger [10] found that there was an average of six authors per paper in the "Original Articles" section of the New England Journal of Medicine. As medicine has become more specialized, cooperation among various subfields has led to a natural increase in the number of authors listed for each article. However, many of today's journals limit the number of authors that can appear, especially for case reports.

Abstract and Key Words

The Abstract is an important section of the manuscript. For many readers, the decision whether or not to read the full text rests on the quality of this summary. Maximum information about the manuscript must be provided in a minimum number of words, since many journals have a set word count for abstracts. The tremendous number of biomedical articles and difficulty selecting the pertinent literature from this pile led to the concept of the "structured abstract." In this type of abstract, the information is divided under subheadings, such as Objective, Clinical Presentation, Intervention, and Conclusion. Such reporting is expected to supply the reader with uniform and standardized information. The argument against this format is that it reduces creativity. Structured abstracts are requested by some journals in order to facilitate more rapid identification of articles, to pave the way for more precise computerized literature searches, and to help editors during the peer review process [1, 12,

23]. The Key Words should appear in alphabetical order, and be in accordance with the medical subject headings list in Index Medicus [13].

The credibility of a manuscript and the size of the audience it reaches are, to some extent, reflected by the number of times it is cited in other articles. The Title, Abstract, and Key Words are all important elements for catching the reader's attention.

Introduction

The Introduction is the first part of the classical IMRAD format. This section familiarizes the reader with the problem, question, or hypothesis that the paper will discuss. The known aspects of the subject are summarized as orientation to the topic. The last portion of this section should tell what is unknown or problematic about the specific subject [8]. The purpose of the study is also conveyed in the Introduction.

Material and Methods

This section describes the entire process that was used in effort to answer the problem described in the Introduction. To start, the author should outline the material and equipment used, and their sources. The study design (descriptive study, randomized clinical trial, prospective or retrospective study) should also be stated. Materials and Methods may be divided into subsections to impart a better understanding of the process that was undertaken.

No results are reported in this section, but the statistical analysis should be explained [8]. Some basic statistical concepts that should be noted include randomization, sample size (intended size), number of treatment groups, conditions for stopping the investigation, the extent of significance testing, and confidence intervals. If a study is randomized, the writer must specify the type of randomization (simple, stratified, blocked, or stratified/blocked). The number of patients to be included in a study is determined before the investigation begins. Statistical power calculations are used for such determinations.

In most studies, two groups are compared with either a placebo group or each other. However, there are statistical methods for comparing more than two groups.

The following are the four statistics-related scenarios for ending a study:

1. When no intended trial size has been specified, the study can be terminated at any time
2. The designated study period ends before the pre-set trial size has been achieved
3. The study is expanded beyond the intended sample size to achieve statistically significant results and is ended when such results are reached
4. If the results achieved at a certain point reveal statistically significant conclusions, then the study may be stopped before the intended sample size is reached.

Significance testing is important for interpreting medical data. Overuse of this should be avoided, and exact P values should be reported instead of stating the P value as greater or less than 0.05. Confidence intervals are closely related to significance testing. They represent the range of values within which the true population probably lies [2, 17, 21].

Results

In this section, the author reports all the concrete evidence. The structure of Results should match that of Material and Methods, in that there are corresponding results for each method that has been described [8]. Results for clinical trials must list the number of subjects who completed the protocol in each group, the total number lost to follow-up, and individuals who were excluded or withdrawn, along with reasons for this. The writer must also report the demographics and clinical characteristics of the study population, the duration of the trial, and any differences between what was planned and what was actually done, including reasons for the differences.

Discussion

The Discussion is the final part of the main text. Answers and solutions to the questions and problems posed in the Introduction are given. This section should include not only the positive and desired or expected results, but should also note any deficiencies, discrepancies, and limitations of the study, including possible sources of bias. The results are compared with findings in the pertinent literature, and similarities and differences are highlighted. For clinical trials, the Discussion must include the extent to which findings and conclusions can be generalized to other populations, with implications for applicability and exclusions [22].

References

All references should be listed either in the order they appear in the text or alphabetically, based on the last name of the first author. The list must be complete and accurate. If the same author has written more than one of the references, then the list order should be based on year of publication, from earliest to most recent. Personal communications and unpublished data may also be referenced. Reporting of personal communication must be kept to a minimum, and should be used only to convey data not found in other sources. The authors must find and read all the references, and cross-referencing should be avoided. Each item in the reference list must be cited in the text.

Types of Manuscripts

Case reports, retrospective-prospective studies, and randomized clinical trials (RCT) [4, 17, 19, 20] are the three major forms of clinical medical writing.

Case Reports

The Case Report is probably the most widely published type of manuscript. The reasons for this may be that it is relatively easy to write this kind of article, and the frequency of interesting isolated cases is high. Regarding format, the manuscript is divided into Introduction, Case Report, and Discussion sections. If extensive laboratory studies are done, then a Material and Methods section should also be included.

There are three types of case presentations. The first is the classical type, in which the findings for only one patient are presented. The second is similar to the classical type, but with two or more similar cases presented. The third form resembles a research paper, and includes personal or institutional series of patients with possible additions from the literature. This third type is more informative, and better characterizes a disorder, especially a rare one. Such reports may also shed light on diagnostic or therapeutic factors that could be used to evaluate a previously undescribed aspect of a disorder. These studies do not necessarily present a specific hypothesis, but they must outline good rationale. Essentially, they seek answers to specific questions [8]. The above-mentioned features are important factors in the evaluation of case studies. Rarity alone is not enough for a Case Report to merit publication.

Retrospective and Prospective Studies

A retrospective study is defined as one conducted after the events under investigation have already occurred. One disadvantage of retrospective studies is that they are sometimes difficult to interpret. Data may be incomplete, missing, or impossible to obtain. A prospective study is one in which the investigation is performed as the events under study occur.

Randomized Clinical Trial

Randomized Clinical Trials are studies in which subjects are randomly assigned to undergo or not undergo an experimental preventive, therapeutic, or diagnostic procedure, and are then followed to determine the effect of the intervention [1]. This is a reliable method for assessing the efficacy of health care interventions. These reports should supply the reader with necessary information about the design, execution, analysis, and interpretation of the trial. Such details help the reader make judgments regarding the internal and external validity of the investigation [4, 9, 19, 20, 23].

To give the reader an idea of the frequency of the different types of manuscripts that are published, the year 2000 issues of the journals *Acta Neurochirurgia*, *Journal of Neurosurgery*, and *Neurosurgery* were reviewed and the results are summarized in Tables 1, 2, and 3, respectively. *Neurosurgery* classifies manuscripts in a more detailed manner, so only the clinical studies and case presentations are listed in Table 3.

The credibility of a printed journal and the respect it holds in the scientific community depend heavily on the evaluation process, editorial work, and peer view prior to publication. Referees and journal editors must remain impartial, and should evaluate manuscripts without making any pre-judgments associated with individuals, institutions, or medical subjects [11, 25].

Table 1. *The Distribution of the Different Types of Manuscripts that Appeared in Acta Neurochirurgia, Year 2000 (Total of 203 Manuscripts)*

	n	%
Clinical study	93	45.81
Case report	60	29.55
Experimental research	23	11.33
Technical note	15	7.38
Historian's corner	6	2.6
Biographical sketch	4	1.9
Cadaveric anatomic study	2	0.985

Table 2. *The Distribution of Different Manuscript Types Published in the Journal of Neurosurgery, Year 2000 (Total of 324 Articles)*

	n	%
Clinical articles	124	38.27
Case reports	73	22.53
Lab. investigations	70	21.6
Case illustration	32	9.87
Technical Note	19	5.86
Historical vignette	5	1.54
Presidential address	1	0.003

Table 3. *The Distribution of Clinical and Case Studies in Neurosurgery, Year 2000 (Total 355 Articles)*

	n	%
Clinical studies	99	27.88
Case reports	61	17.18
Technical case reports	12	3.38
Case problems	3	0.01

A journal submission must be of high quality in order to be accepted. One main way to improve manuscript quality is to foster a stronger educational focus on scientific writing skills. This type of teaching needs to be part of basic neurosurgical training. A good example is the program in the Neurosurgery Department at the University of California, where residents are supplied with a publication kit and attend a series of seminars and one-to-one tutorials with academic biomedical editors [8].

In summary, the task of preparing a manuscript for a printed journal is not an easy one. Aside from the actual writing, the main difficulty is in presenting completely original information, which is what most publications require. Although the preparation process can be challenging, the abundance of biomedical journals and the demand for manuscripts stimulates writing and creates a platform on which authors can present their work. Regardless of the problems that may be encountered in the writing, submitting, and publishing stages, authors are ultimately responsible for ensuring that the work they present is accurate, scientific, honest, and unbiased. This is the only way to achieve and maintain the high level of scientific quality we all seek.

References

1. Ad Hoc Working Group for Critical Appraisal of the Medical Literature (1987) A proposal for more informative abstracts of clinical articles. Ann Int Med 106: 598–604
2. Altman DG, Doré CJ (1990) Randomisation and baseline comparisons in clinical trials. Lancet 335: 149–153
3. Apuzzo MLJ, Hodge CJ (2000) The metamorphosis of communication, the knowledge revolution, and the maintenance of a contemporary perspective during the 21st century. Neurosurgery 46(1): 7–15
4. Begg C, Cho M, Eastwood S, Horton R, Moher D, Olkin I, Pitkin R, Rennie D, Schulz KF, Simel D, Stroup DF (1996) Improving the quality of reporting of randomized controlled trials. The CONSORT Statement. JAMA 276(8): 637–639
5. Bhopal R, Rankin J, McColl E, Thomas L, Kaner E, Stacy R, Pearson P, Vernon B, Rodgers H (1997) The vexed question of authorship: views of researchers in a British medical faculty. BMJ 314: 1009
6. Billing JS (1881) An address on our medical journal. Br Med J 2: 262–268
7. Durack DT (1978) The weight of medical knowledge. N Eng J Med 298: 773–775
8. Eastwood S, Derish PA, Berger MS (2000) Biomedical publication for neurosurgery residents: a program and guide. Neurosurgery 47(3): 739–749
9. Freiman JA, Chalmers TC, Smith H Jr, Kuebler RR (1978) The importance of beta, the type II error, and sample size in the design and interpretation of the randomized controlled trial: survey of 71 'negative' trials. N Eng J Med 299: 690–694
10. Friesinger GC (1986) Who should be an author? J Am Coll Cardiol 8: 1240–1242
11. Fye WB (1990) Medical authorship: traditions, trends, and tribulations. Ann Int Med 113(4): 317–325
12. Haynes RB, Mulrow CD, Huth EJ, Altman DG, Gardner MJ (1990) More informative abstracts revisited. Ann Intern Med 113: 69–76
13. Hoen WP, Walvoort HC, Overbeke AJ (1996) What are the factors determining authorship and the order of the authors' names? JAMA 280(3): 217–218
14. Holmes LS, Grayson TH (1999) Neurosurgery online and the internet: prospects for the next decade. Neurosurgery 44(1): 9–11
15. International Committee of Medical Journal Editors (1985) Guidelines on authorship BMJ 291: 721
16. International Committee of Medical Journal Editors (1997) Uniform requirements for manuscripts submitted to biomedical journals. Ann Int Med 126(1): 36–45
17. Moher D, Dulberg CS, Wells GA (1994) Statistical power, sample size, and their reporting in randomized controlled trials. JAMA 272: 122–124
18. Pareras LG, Martin-Rodriguez JG (1996) Neurosurgery and the internet: a critical analysis and a review of available resources. Neurosurgery 39(1): 216–233
19. Peto R, Pike MC, Armitage P, Breslow NE, Cox DR, Howard SV, Mantel N, McPherson K, Peto J, Smith PG (1976) Design and analysis of randomized clinical trials requiring prolonged observation of each patient. I. introduction and design. Br J Cancer 34: 585–612
20. Peto R, Pike MC, Armitage P, Breslow NE, Cox DR, Howard SV, Mantel N, McPherson K, Peto J, Smith PG (1977) Design and analysis of randomized clinical trials requiring prolonged observation of each patient. II. analysis and examples. Br J Cancer 35: 1–38
21. Pocock SJ, Hughes MD, Lee RJ (1987) Statistical problems in the reporting of clinical trials. A survey of three medical journals. N Eng J Med 317(7): 426–432
22. The Asilomar Working Group on Recommendations for Reporting of Clinical Trials in the Biochemical Literature (1996) Checklist of information for inclusion in reports of clinical trials. Ann Intern Med 124: 741–743

23. The Standards of Reporting Trials Group (1994) A proposal for structured reporting of randomized controlled trials. JAMA 272(24): 1926–1931
24. Wilkins RH (1981) Preparation of a neurosurgical manuscript, with emphasis on library research. Clin Neurosurg 28: 173–192
25. Working Group on Recommendations for Reporting of Clinical Trials in the Biomedical Literature (1994) Call for comments on a proposal to improve reporting of clinical trials in the biomedical literature. Ann Int Med 121(11): 894–895

Correspondence: N. Altınörs, Department of Neurosurgery, Başkent University, 10 Sokak, No. 45, Bahçelievler, 06490 Ankara, Turkey.

Acta Neurochir (2002) [Suppl] 83: 121–125

Appropriate Analysis and Presentation of Data is a must for Good Clinical Practice

M. Hayran

OMEGA Contract Research Organization, Ankara, Turkey

Summary

Good Clinical Practice (GCP) is defined as an international ethical and scientific quality standard for designing, conducting, monitoring, auditing, analyzing and reporting trials that involve the participation of human subjects. This paper focuses mainly on the issues that need attention at the time of statistical analysis and reporting of results. Findings from a review of published articles in Turkey are also presented. The types of variables, the distributions of variables, the number of groups compared, the dependency structure among these groups and the primary goal of the analysis determine the appropriate method to be selected for the statistical analysis. The review of a stratified sample of research articles from 60 journals published in 1992 in Turkey revealed that in 56% of the cases the statistical methods were improper or inadequate. In 15% of the articles the authors failed to select an appropriate design for the proposed aim mentioned in the manuscript. Despite the recent improvements, the necessity and the value of performing and presenting research according to the international standards remains to be assimilated better by Turkish investigators.

Keywords: Clinical trials; GCP; data analysis; manuscript preparation.

Introduction

Good Clinical Practice (GCP) is defined as an international ethical and scientific quality standard for designing, conducting, monitoring, auditing, analyzing and reporting trials that involve the participation of human subjects. This standard aims to provide public assurance that the rights, safety and well-being of trial subjects are protected, and that the clinical trial data are credible [5].

There are some essential steps that need to be followed throughout a medical research. These steps are the design, conduct, analysis, publication, interpretation and application. For any research to make a useful contribution to the pool of scientific knowledge, the investigators must make sure they have spent enough time at each of the steps so that they meet the scientific standards. In this paper, we will mainly focus on the statistical analysis phase, which is the third step mentioned above. However, in this paradigm, for any of the phases to make scientific sense, the preceding steps need to be performed appropriately, so we will include in our discussion the design and conduct phases.

Design Phase

For the data collected in the trial to be valid and reliable, the design must be appropriate and statistically justified. It is not ethical to involve human subjects in trials where the design is not up to scientific standards. The statistical plan that will be used at the time of analyses must be decided upon before the trial is started and must be identified in the research protocol. This requires that the primary and secondary outcomes as well as the eligibility criteria of the analysis populations be determined explicitly. Along with the statistical plan, these design issues need to be mentioned in the appropriate sections of the protocol.

Conduct Phase

The observations and the findings should be recorded on the case report forms and they must be signed. The outputs from the computer printouts need to be dated and signed. The data entry and editing must be performed only by the data management personnel and must not be intervened by anyone else in the investigation staff. The monitoring and auditing procedures must be performed in a timely manner and the reports from these monitoring and auditing visits should be evaluated by the appropriate institutions to make sure the quality of the trial is not compromised at any stage during the trial. Source documents and other relevant documentation must be stored for several years following termination of the study.

Analysis Phase

Preparation

One very important step is to organize the study data so as to bring order to the several forms of information (such as notes, test results, report forms, etc.) collected during research before beginning to analyze and interpret them [2]. Any operation on the trial data must be performed by persons with appropriate education and experience using valid computer software that are known to be error free. Missing data and illogical data must be handled appropriately. The confidentiality of the individuals must be assured in any version of electronic data and personal identifiers must be stripped off. Pseudo-identifiers that the initial central data manager can decode may be used so that referrals can be made back to the source documents, if there is a need to verify the data.

As was mentioned earlier, the statistical procedures must be those that are identified in the research protocol. If statistical methods differing from the initially identified ones are to be used, the reasons for this modification must be mentioned in the analysis report. The plan and conduct of the analysis must be carried out or verified by experienced specialists in the field of biostatistics. To decide on the appropriate analytic method one must give answers to the following questions:

- What is the type of each variable in the data?
- Are the numerical variables normally distributed?
- How many analysis groups are present in the trial?
- Are the groups of the data dependent or independent of one another?
- Is the primary goal the comparison of groups or the prediction or estimation of several values or variables?

Types of Data

The data types can be mainly divided into two groups as categorical and numeric. The numeric variables conform to the arithmetical rules, i.e. mathematical operations such as addition, multiplication, division, subtraction, can be performed on these variables (e.g. age, serum Ca level, etc). On the other hand, the categorical variables do not carry this property. Categorical variables may be further divided into nominal (simple categorical), dichotomous and ordinal variables. The ordinal variables have coded values that reflect the superiority and inferiority among each other (e.g. stage, grade). We know that a value of 1 is higher than 0 and less than 2 but a mathematical procedure will not make sense (e.g. 3 minus 2 is not equal to 2 minus 1). In nominal variables this is not the case. The codes are just to identify a group but not to identify which group is superior (e.g. eye color, marital status). If in the data, we coded 'blue eyes' as 1 and 'brown eyes' as 2, that will not mean that having brown eyes is superior to having blue eyes. The main characteristic of dichotomous variables is that it can take on only two values. They may resemble the nominal variables (sex; 1: male-0: female) or ordinal variables (fever; 1: present-0: absent) in terms of comparability of the coded values.

Normal Distribution

The numerical variables follow a predefined pattern in terms of probability to see each value for that variable in the general population. The higher the frequency of a value the more the chance to come across such a value in the population. As an example, the chance to meet a 95 year old person on the streets is a lot less than the chance of seeing a 10 year old. The charts showing the frequency of values of a variable is called a histogram and gives us information about the distribution of a variable.

The normal distribution can be visually verified by looking at a histogram. In the histogram of a normally distributed variable the frequency peaks around the mean value and the median and the mode values of a variable are very close to the peak (they are actually equal to the mean for the perfectly normally distributed variable) as well. The normal distribution histogram is symmetric and there are equal number of observations above and below the mean value. The distributions that are not normal are not symmetric, the mean and median values are quite different from each other and the distribution is skewed to either right or left.

The justification that a distribution is normal can be made by performing statistical tests. As a general rule the variables that have a standard deviation less than 30% of its mean tend to be normally distributed.

Summarizing Data

The appropriate summary figure to be reported for different types of variables are different. For nominal variables the proportion of cases for each value needs to be reported. This may be used for the ordinal variables as well. Another summary measure useful to report for ordinal variables is the median value. The median value is used to define the central tendency of a non-normal distribution for numeric values as well. For normally distributed values the mean value needs to be reported.

The central tendency measure gives the reader information about the peak of a frequency distribution for a numerical variable. There is another measure which is the dispersion measure that gives us information about how further away the values are from one another. For normally distributed numeric variables this is given by reporting the standard deviation. For ordinal variables and for those that are not normally distributed, the range (the difference between minimum and maximum of a variable) or the interquartile range (the difference between 25th and 75th percentiles of a variable) should be reported.

Dependent and Independent Data Groups

If the data is grouped in a way that the values for each group are obtained from people or subjects that are different from the other group, then these groups are independent. If values obtained from the same person or same subject are split into different groups, then these groups would be dependent groups.

Which Statistical Method to Choose

If the variable type is simply categorical, than the cross-tab techniques are usually used for comparisons. For ordinal variables and numeric variables without normal distribution nonparametric tests are used. Only numeric variables that conform to a normal distribution can be analyzed with the parametric tests. See Table 1 for a list of appropriate tests to be performed for the given scenarios [1, 4, 6–10].

Table 1. *Selection of Appropriate Statistical Tests in Several Situations Comparing Outcomes of Different Groups*

When comparing	Of	That are	Use
Proportions	2 groups	independent	Chi-square or Fisher tests
Proportions	more than 2 groups	independent	Chi-square test
Proportions	2 groups	dependent	Mc Nemar test
Proportions	more than 2 groups	dependent	Cochran's Q test
Means*	2 groups	independent	Student's t test
Means*	more than 2 groups	independent	ANOVA test
Means*	2 groups	dependent	Student's paired t test
Means*	more than 2 groups	dependent	repeated measures ANOVA test
Medians	2 groups	independent	Mann-Whitney U test
Medians	more than 2 groups	independent	Kruskal-Wallis test
Medians	2 groups	dependent	Wilcoxon test
Medians	more than 2 groups	dependent	Friedman test

* Don't compare means if the numerical variable is not normally distributed, compare medians instead.

Predicting Values or Estimating Magnitude of Associations

Usually the predictions are made using regression tests and the basic associations are analyzed using correlation tests. However, extreme caution needs to be exercised when selecting the correct tests and making sure the assumptions for each particular test are not violated in such a way that the results become invalid altogether.

Remember once again that at the minimum, the supervision of a specialist in biostatistics is necessary when choosing and performing the appropriate statistical method.

Quality of Published Results of Trials

Publishing Findings

The interpretation and publication of the results are of course essential steps for the information from the trial to be disseminated to the relevant audience. This would help the professionals in the corresponding field to implement the findings in their own work and further develop and test new hypotheses. Once again,

Table 2. *Distribution of the Study Types in Articles Published in Turkey*

Type of study	Turkey		Surgery		Neurosurgery	
	Count	Percent	Count	Percent	Count	Percent
Case report	112	22%	80	29%	15	44%
Other descriptive	147	28%	71	25%	9	26%
Intervention trial	130	25%	80	29%	6	18%
Cross-sectional	73	14%	22	8%		0%
Methodological	24	5%	10	4%	2	6%
Case-control	18	3%	12	4%	2	6%
Cohort	8	2%	4	1%		0%
Total	517	100%	279	100%	34	100%

Table 3. *Accuracy in Identifying the Study Type by the Authors in Published Articles in Turkey*

Type of Study	Turkey		Surgery		Neurosurgery	
	Count	Percent	Count	Percent	Count	Percent
Mentioned correctly	47	12%	37	19%	1	5%
Mentioned incorrectly	34	8%	10	5%	–	–
Not mentioned in article	324	80%	152	76%	18	95%
Total	405	100%	199	100%	19	100%

such publications will only be useful if they base upon well-designed, well-conducted, unbiased and properly analyzed research [3].

Quality of Published Research in Turkey

In an earlier effort which aimed at estimating the quality of trials reported in the published research articles in Turkey, a stratified sample from 1250 manuscripts published in 60 journals in 1992 were analyzed. Five hundred seventeen publications constituted the sample. Two independent, experienced reviewers rated each manuscript for various aspects of the appropriateness of their design, conduct, analysis, and presentation of the results.

It was observed that 50% of the articles were either case reports or other descriptive studies (Table 2). Unfortunately, only 12% of the study type was correctly mentioned in the manuscript by the authors while in 88% it was either omitted or mentioned incorrectly (Table 3). The selected study design was appropriate in 85% of the cases for the aim identified by the authors in the manuscript (Table 4).

Table 4. *Appropriateness of the Study Design for the Aim Mentioned in the Manuscript in Published Articles in Turkey*

Study design	Turkey		Surgery		Neurosurgery	
	Count	Percent	Count	Percent	Count	Percent
Appropriate for mentioned aim	442	85%	212	95%	17	89%
Not appropriate for mentioned aim	75	15%	10	5%	2	11%
Total	517	100%	222	100%	19	100%

Table 5. *Presence of an Explicitly Mentioned Hypothesis in Published Trials in Turkey*

Hypothesis	Turkey		Surgery		Neurosurgery	
	Count	Percent	Count	Percent	Count	Percent
Mentioned in the article	28	13%	6	7%	–	–
Not mentioned in the article	191	87%	85	93%	10	100%
Total	219	100%	91	100%	10	100%

Table 6. *Presence of an Explicitly Mentioned Sample Size in Published Trials in Turkey*

Sample size	Turkey		Surgery		Neurosurgery	
	Count	Percent	Count	Percent	Count	Percent
Given in the article	4	2%	2	2%	–	–
Not given in the article	253	98%	85	98%	10	100%
Total	257	100%	87	100%	10	100%

However, for the articles reporting trials, in 87% of the cases the authors failed to mention the hypothesis which was tested by the reported trial (Table 5). Moreover, 98% of the articles that needed to report their sample size failed to do so (Table 6). In 56% percent of the cases, the statistical analyses were far from being adequate for that particular research, either lacking appropriate analytic approaches or choosing techniques that were incompatible with the data.

One should keep in mind that although these results reflect findings from the last decade and although

Table 7. *Appropriateness of the Selected Statistical Method in Published Trials in Turkey*

Statistical method	Turkey		Surgery		Neurosurgery	
	Count	Percent	Count	Percent	Count	Percent
Very good/good/fair	125	44%	37	30%	3	30%
Bad/very bad	157	56%	88	70%	7	70%
Total	282	100%	125	100%	10	100%

the design, conduct, analysis and presentation of trials have become more scientific in the last couple of years in Turkey, there is still an increasing need to make our investigators understand better the necessity and the value of performing and presenting research according to the international standards.

References

1. Cochran WG (1950) The comparison of percentages in matched samples. Biometrika 37: 256–266
2. Gay LR, Airasian P (2000) Qualitative research: data analysis. In: Davis KM (ed) Educational research, competencies for analysis and application, 6th edn. Prentice Hall, New Jersey, p 237
3. Jadad A (1998) Reporting and interpreting individual trials: the essentials. In: Randomised Controlled Trials, 1st edn. BMJ Books, London, pp 61–77
4. Friedman M (1937) The use of ranks to avoid the assumption of normality implicit in the analysis of variance. J Am Statist Assoc 32: 675–701
5. International Conference on Harmonisation Expert Working Group (1998) ICH Guideline for good clinical practice & declaration of Helsinki, 2nd edn. Janssen Research Foundation, Belgium, p 1
6. Kruskal WH, Wallis WA (1952) Use of ranks in one-criterion variance analysis. J Am Statist Assoc 47: 583–621
7. Mann HB, Whitney DR (1947) On a test of whether one of two random variables is stochastically larger than the other. Annals of Math Stat 18: 50–60
8. Muller KE, Curtis NB (1989) Approximate power for repeated-measures ANOVA lacking sphericity. J Am Statist Assoc 84: 549–555
9. Rosner B (2000) Fundamentals of Biostatistics, 5th edn. Duxbury Press, Pacific Grove, California
10. Wilcoxon F (1945) Individual comparisons by ranking methods. Biometrics 1: 80–83

Correspondence: Murat Hayran, OMEGA Contract Research Organization, Güniz Sok. No: 32/12, Kavaklıdere, Ankara, Turkey.

Acta Neurochir (2002) [Suppl] 83: 127–131

Why Collect Medical Books?

H. August van Alphen

Department of Neurosurgery, Academisch Ziekenhuis Vrije Universiteit, Amsterdam, The Netherlands

Summary

The study of the original sources in the medical literature is an essential part and condition for doing good medical research. The love of books is conducive to a proper use of the source material. To support this statement, a short description of some famous book collectors in medical history is presented in this article.

Keywords: Medical books; medical book collectors; history of medicine.

In 1995, the US literary editor and columnist Nicholas Basbanes, a dedicated bibliophile himself, published a book on book collecting and book collectors [2]. He spent eight years conducting research, visiting numerous libraries all over the country, following leads and interviewing hundreds of people for this book. He concluded that, indeed, passionate book collecting remains the only hobby to have a disease named after it, which is: A gentle madness. This is also the title of the book. In this book, he describes, among others, the story of Stephen Blumberg from Ottumwa, Iowa and his peculiar book collection. Blumberg appeared to be the most enterprising bibliomaniac and biblioklept of the last century. In the period between 1970 and 1990 he had stolen about 23.000 books form 268 libraries in fifty-five states, two Canadian provinces, and the district of Columbia. Among those libraries were that of the University of California at Los Angeles, University of California at Riverside, Duke University, Harvard University, the Universities of Minnesota and Cincinnati, New Mexico and Oregon, Washington State University, University of Southern California etc. When he went to a library, mostly with a stolen identity card, he knew exactly what he was looking for. His main interest was Americana and Incunabula. He knew about a book before he acquired it and he always tried to fill the gaps that still remained in his growing collection. Before leaving the library, he removed most traces of identification from the volumes. But he kept the bookplates in several scrapbooks as a little collection of mementos in its own right. The total value of his ultimate collection was estimated some 20 million US dollar. The most remarkable fact was, that these robberies remained largely unnoticed.

When Blumberg was arrested finally in March 1990, it was not because of any alarm that had been sounded throughout the country. No, it was because a friend of him of fifteen years turned him in for a 56.000 US dollar bounty negotiated with the Justice Department.

After his capture, the police rented a forty-foot tractor-tailor to bring the nineteen tons of books in eight hundred and seventy-nine cardboard packing boxes to a secret warehouse in Nebraska, where nine rooms of the building were stacked top to bottom with books. The problem was, that no-one had an earthly idea where any of them came from. And making the dilemma doubly difficult was that very few of the books had ever been known to be missing in the respective institutions.

During trial, Blumberg told the judges that he never took a book with the intention of selling it or getting any the richer from it. He had the idea to keep them all with him. He considered his actions as a form of interlibrary loan.

One may doubt as to whether you can call this a gentle madness. But, fortunately, most book collectors are not that extreme as Stephen Blumberg. What, on the other hand, the underlying motive for the passion to collect and to possess books for those people is, is not very clear. Restricting ourselves to the medical world, there are many famous medical doctors, who also are or were passionate book collectors. The following will present a look at some of them.

Sir William Osler (1849–1919) was born in Upper Canada, and had his medical training at the Toronto Medical School and McGill Medical School in Montreal. Later, he became professor of Medicine in McGill University, professor of Clinical Medicine in the University of Pennsylvania in Philadelphia, professor of Medicine in the Johns Hopkins University in Baltimore and finally professor of Internal Medicine in the University of Oxford, England.

Osler loved reading books and he knew a lot about medical libraries. Once in an address, he emphasised the importance of reading as a part of post-graduate study and he said that there had been men whose only book was nature, but they were the exceptions. The average non-reading doctor might play a good game of golf or of bridge, but professionally he was a lost soul (4, p 184).

In another address he spoke about the use of a great medical library for teachers, for general practitioners, and for another group to which he himself belonged. He said (3, p 545): "There is a third class of men in the profession to whom books are dearer than to teachers or practitioners – a small, a silent band, but in reality the leaven of the whole lump. The profane call them bibliomaniacs, and in truth they are at time irresponsible and do not always know the difference between meum and tuum. I dare not further characterise them. Loving books partly for their contents, partly for the sake of the authors, they not alone keep alive the sentiment of historical continuity in the profession, but they are the men who make possible such gatherings as the one we are enjoying this evening (which was the Boston Medical Library). We need more men of their class, particularly in this country, where everyone carries in his pocket the tape-measure of utility."

It is not known whether this was the real reason or motive for William Osler to gather books himself. He finally possessed a fabulous library of almost 8000 volumes, which he bequeathed to the McGill Medical Library in Montreal [10].

Osler's pupil and later dear friend and biographer [3, 4], *Harvey Cushing* (1869–1939), probably adopted the passion for books from him. Born in Cleveland as the son of a family doctor, who also used to be a book collector, Harvey studied medicine at Harvard Medical School in Boston. After finishing medical school, William Halsted summoned him to Baltimore to train as a general surgeon. Here, Cushing soon developed a close friendship with William Osler, who urged him to apply himself to the surgical management of disorders of the nervous system. And so, from the beginning of 1898, Cushing started to focus on the pathology of the nervous system. After finishing his surgical training in 1900, Cushing spent a "Wanderjahr" in Europe on the advice of William Welch, pathologist at The Johns Hopkins, Osler and his father. During this period Cushing visited the anatomical theatre at Padua, Italy, where Andreas Vesalius had performed the public dissections pictured in the frontispiece of his book: De Humani Corporis Fabrica [14]. Cushing was excited about it. Three years later, he wrote to his father (7, p 230): "Dr. Osler has started me on a Vesalius essay. He has turned over to me pro tempore a stunning copy of the "De Corpora Humanis Fabrica" (Cushing even didn't know the exact title of the book at that time) with the famous plates etc. I want very much to collect photographs of the various portraits and as many engravings of Vesalius himself as possible so if you run across any of them in your perusal of catalogues or see a notice of the sales of any of his books I wish you would let me know." The paper on which Osler had started Cushing was given the title "The Books of Vesalius" and presented before the Book and Journal Club of the University of Maryland. The day after that meeting, someone left on Cushing's doorstep a handsome copy of the second edition of the Fabrica [7]. And that was the beginning of another fabulous library, and, in addition, the greatest existing Vesaliana collection.

Cushing's collection, also comprising some 8.000 books, was quite different from Osler's and had a more personal character. Especially his Vesaliana was unique. The last book Cushing wrote was: A Bio-bibliography of Andreas Vesalius [5], which was published after his death by his student and friend John Fulton in 1943, the year of the 400th anniversary of the Fabrica.

In 1934, five years before his death, Cushing decided to leave his whole collection to the Yale Medical Library [6, 11]. It was John Fulton who was largely instrumental in bringing his former teacher to Yale as Sterling Professor of Neurology after his retirement as a neurosurgeon a year before. Later, Cushing became the Director of Studies in the History of Medicine at Yale.

John Fulton (1899–1960), it is not at all surprising, was another ardent book collector. Fulton was Stirling Professor of Physiology in the Yale University School of Medicine. He was Cushing's biographer, which biography was published in 1946 [7]. In 1951 Fulton resigned as Professor of Physiology to become Sterling Professor of the History of Medicine and Chairman of the newly created Department of the History of Medicine at Yale. He thrust himself energetically into his new endeavour and soon his department became one of the foremost centers in the world for studies of medical history. And he became editor of the journal of the History of Medicine and Allied Sciences. On the occasion of his sixtieth birthday his friends presented to him a booklet, titled: The Making of a Library [1], and containing extracts from letters between Harvey Cushing, Arnold Klebs and himself, in which they discussed the plans to make a Klebs-Fulton-Cushing collection of medical books and to leave their combined libraries to the Yale Medical Library. After this decision was made, they began to collect systematically, each one in a few specific fields so that the collections, when brought together, might have unity, and overlapping and duplication might be avoided.

Another student of Cushing's was the future neuropathologist *Cyril Brian Courville* (1900–1968). He was born in Michigan, and graduated in 1925 from what is now Loma Linda University School of Medicine in California. In 1927–1928, he spent two years as volunteer assistant to Cushing at the Peter Bent Brigham Hospital in Boston. Courville became Professor of Nervous Diseases at Loma Linda in 1933, and Director of the Santiago Ramon y Cajal Laboratory of Neuropathology at the Los Angeles County Hospital in 1934.

It is not sure when he started collecting books, but without any doubt, his working with Cushing must have been a great stimulus to him. He brought together books on the various aspects of the neurosciences and related topics, representing his neurological interests during his long and accomplished career in neuropathology. Finally, he left one of the most extensive collections in this area, together with a collection of books on historical medicine outside the neurosciences, in total some 3.000 volumes, to the Medical Sciences Library of the University of California at Irvine [13].

The greatest collection of old medical books and works on the history of medicine is unmistakably that of the Swedish general surgeon, *Erik Waller* (1875–1955). He was born at Önum in the province of Västergötland in western Sweden, and took his M.B. degree at Uppsala University. He had his surgical training with the famous surgeon John Berg at Karolinska Institute in Stockholm. After completion of his studies, he worked as a surgeon in different parts of the country. From 1940 to 1946 he was librarian of the Swedish Medical Association in Stockholm.

Waller was internationally known as one of the most prominent book collectors of our days. It is hardly an exaggeration to say that his collection, in the fields it covers, is unparalled in the last century. It comprises practically all major works published in the realm of medicine before 1800, as well as a highly representative selection of later literature on the subject. While the collections of Osler and Cushing, as mentioned before, each comprises some 8.000 items, Waller's includes nearly 21.000, and it is also far superior in quality. This library had been collected by him book by book. He started collecting in the early 1900's. At first, manuscripts and autographs were his main interest; he collected nearly 20.000 letters, dating from the beginning of the 16th to the early 20th century. A great deal of these letters were written by or are addressed to men famous in the history of medicine. Later, Waller started to collect books, which became his main achievement.

And no wonder, during his whole career he maintained stimulating and close contacts with scholarly collectors as Harvey Cushing, Arnold Klebs and John Fulton. Cushing visited Waller in 1927 and wrote to Arnold Klebs (7, pp 579–580): "I am this far on the way. A most delightful spot and this man Waller has such books as you never saw!!! I have almost decided to remain here a full month someday – perhaps with you." Waller's own writings include a study, published in 1936, of the unique edition of Vesalius' De Humani corporis fabrica librorum epitome, which is the abbreviated edition of Vesalius' main work [15] and a biography of his close friend Arnold Klebs [16]. It was Waller's great desire to make his collections as useful as possible by making them available to scholars. Therefore, he decided to give his whole library to the University of Uppsala, Sweden [12].

Waller brings us to the last medical book collector, whom I want to mention: *Arnold C. Klebs* (1870–1943), a Swiss tuberculosis specialist from Berne. Klebs became very close with Cushing and also Osler's influence on him was apparent [16].

In 1895 Klebs' father came from Switzerland to the United States as a pathologist to a tuberculosis sanatorium in Asheville, North Carolina. His association there was short-lived, for in 1896 he became Professor of Pathology at Rush Medical College in Chicago. Arnold Klebs followed his father to the United States in 1896, and after a year with Osler at The Johns Hopkins in Baltimore he began to specialise in tuberculosis and then, he became head of a sanatorium at Citronelle, Alabama. Later, he settled in Chicago where he practised for some ten years as a tuberculosis specialist. He directed the Chicago Tuberculosis Institute and became one of the first directors of the National Tuberculosis Association. In 1909, at the request of Osler, he edited a 939-pages volume on tuberculosis with chapters from Osler and several others. In 1910 he retired from practice to return to his native Switzerland, where he lived until his death in March 1943.

In 1896, when Klebs first came to the United States, he began, under William Osler's influence, to collect the literature of his specialty, and at the time of his death he had accumulated nearly 3000 items bearing on the history of tuberculosis alone.

The earliest mention of a meeting between Harvey Cushing and Arnold Klebs is noted in a letter from Cushing to his father in Cleveland the 2nd of February 1906, in which he wrote (7, p 247) : "There is a nice man named Klebs, almost too nice to live in Chicago, who has been down here (in Baltimore) on an occasional visit, and who turned up again a day or two ago to say how-do-you-do to us and good-bye to Dr. Osler."

Klebs and Cushing seemed so different by temperament that many, who knew them, have expressed surprise that they ever should have become so intimate.

So, Klebs was a great friend of Osler, Cushing and William Welch, pathologist at The Johns Hopkins, and in his later years of John Fulton. In 1931 Klebs met Harvey Cushing again in Berne at the occasion of the First International Congress of Neurology. Many years later, John Fulton wrote a historical paper on this meeting, titled: Arnold Klebs and Harvey Cushing at the 1st International Neurological Congress at Berne in 1931, mentioning the reunion of many friends from Europe and the United States. Klebs gave a great dinner-party for all of them [8]. In 1934, Harvey Cushing wrote a letter to Arnold Klebs, telling him the plans he and John Fulton had on their libraries after their death, and asking Klebs: (1, 7, p 647) "I don't know what your own plans may be if you have any. I know that you once thought of establishing at Les Terrasses a foundation for medico-historical studies. This you may still intend to do but if not and if this other idea has any interest for you do let me know." Klebs was a little taken aback by this cool suggestion concerning the disposition of his library, but after discussing many aspects of the disposal he responded favourably. He wrote to Cushing (1, 7, p 647): "But then I think of . . . my own books of which with all my omnivorousness I can but digest a very small number. They get additional value when John blows in and spends a week over some items that I hardly have looked at. Would he like to do it as much if I were not there? Your books at Whitney Avenue last winter had a different sense with you at the hospital accessible for their discussion and the hope that we could go over them together when you were up again. The fun of mere acquisition may be great, and indeed it is a necessary phase, but the working with the books, the discussion about them, and the comparison is, after all, what really is most worth while, don't you think so?"

Now, this line of thought of Klebs carries, in my opinion, the heart of the matter. When we overlook these individuals with their passion for collecting books and with their interrelationship, we must conclude that, in fact, there is no "why" in collecting books. It is just an infectious disease, which only gets sense if you are actively dealing with it and can share the fun with other sufferers from the syndrome. Then, the disease takes a turn for the better and the patient can profit from it. Then you will get a wider view on medicine as a whole or on the specialty your collection is focused on. Like William Osler, who, at the end of the nineteenth century, was one of the last persons to have an overall view of the entire field of medicine, judging by his book, The Principles and Practice of Medicine [9], published in 1892, which has served as the golden standard of medicine for more than 30 years, despite the origin of medical specialties in the same period of time, and which has been considered one of the most influential textbooks of general medicine ever written.

Like Harvey Cushing and John Fulton, who understood very well that a true scientific attitude in doing medical research has to be based upon the sources in the literature, and who stood head and shoulders above their own specialties.

Then, you can say with Sir Winston Churchill, when addressing the Royal College of Physicians in 1944: "The longer you can look back, the further you can look forward."

References

1. Anonymus (1959) The making of a library. Extracts from letters 1934–1941 of Harvey Cushing, Arnold C Klebs, John Fulton. Yale University, New Haven, Connecticut
2. Basbanes NA (1995) A gentle madness. Henry Holt and Company, New York
3. Cushing H (1925) The life of Sir William Osler, vol I. At the Clarendon Press, Oxford
4. Cushing H (1925) The life of Sir William Osler, vol II. At the Clarendon Press, Oxford
5. Cushing H (1943) A bio-bibliography of Andreas Vasalius. Schuman's, New York
6. Cushing H (1943) The Harvey Cushing collection of books and manuscripts. Schuman's, New York
7. Fulton JF (1946) Harvey Cushing. A biography. Charles C Thomas, Springfield Illinois
8. Fulton JF (1941) Arnold Klebs and Harvey Cushing at the 1st International Neurological Congress at Berne in 1931. No place (bound xerox copy from the library of L van Bogaert, Antwerp)
9. Osler W (1892) The principles and practice of medicine, designed for the use of practitioners and students of medicine. Young J Pentland, Edinburgh & London
10. Osler W (1929) Bibliotheca Osleriana. A catalogue of books illustrating the history of Medicine and science collected, arranged, and annotated by Sir William Osler, Bt and bequeathed to McGill University. At the Clarendon Press, Oxford
11. Reeves DL (1963) The Harvey Cushing Library. J Neurosurg 20: 545–556
12. Sallander H (1955) Bibliotheca Walleriana. A catalogue of the Erik Waller collection. Almqvist & Wiksell, Stockholm
13. Thomas MLM and Cooper WC (1978) Catalog of works in the neurological sciences collected by Cyril Brian Courville, MD representative of clinical neurology, neuroanatomy and neuropathology with particular reference to head trauma. Books from Historical Medicine. University of California, Irvine, California
14. Vesalius A (1543) De humani corporis fabrica libri septem. Oporinus, Basel
15. Waller E (1936) Eine unbekannte Ausgabe von Vesals Epitome. Lychnos, Uppsala
16. Waller E (1944) Arnold C Klebs 1870–1943. Lychnos, Uppsala

Correspondence: H. August van Alphen, M.D., Ph.D., Hogerlustlaan 57, 1191 CL Ouderkerk a/d Amstel, The Netherlands.

Author Index

Index of Keywords

Springer-Verlag and the Environment

WE AT SPRINGER-VERLAG FIRMLY BELIEVE THAT AN international science publisher has a special obligation to the environment, and our corporate policies consistently reflect this conviction.

WE ALSO EXPECT OUR BUSINESS PARTNERS – PRINTERS, paper mills, packaging manufacturers, etc. – to commit themselves to using environmentally friendly materials and production processes.

THE PAPER IN THIS BOOK IS MADE FROM NO-CHLORINE pulp and is acid free, in conformance with international standards for paper permanency.